CALCULUS CALCULATOR

CALCULUS CALCULATOR
by David Meredith

Prentice-Hall

YOU SHOULD CAREFULLY READ THE FOLLOWING TERMS AND CONDITIONS BEFORE OPENING THIS DISKETTE PACKAGE. OPENING THIS DISKETTE PACKAGE INDICATES YOUR ACCEPTANCE OF THESE TERMS AND CONDITIONS. IF YOU DO NOT AGREE WITH THEM, YOU SHOULD PROMPTLY RETURN THE PACKAGE UNOPENED.

IT IS A VIOLATION OF COPYRIGHT LAW TO MAKE A COPY OF THE ACCOMPANYING SOFTWARE EXCEPT FOR BACKUP PURPOSES TO GUARD AGAINST ACCIDENTAL LOSS OR DAMAGE.

Prentice-Hall, Inc. provides this program and licenses its use. You assume responsibility for the selection of the program to achieve your intended results, and for the installation, use, and results obtained from the program. This license extends only to use of the program in the United States or countries in which the program is marketed by duly authorized distributors.

LICENSE

You may:

a. use the program;
b. modify the program and/or merge it into another program in support of your use of the program.

LIMITED WARRANTY

THE PROGRAM IS PROVIDED "AS IS" WITHOUT WARRANTY OF ANY KIND, EITHER EXPRESSED OR IMPLIED, INCLUDING, BUT NOT LIMITED TO, THE IMPLIED WARRANTIES OF MERCHANTABILITY AND FITNESS FOR A PARTICULAR PURPOSE. THE ENTIRE RISK AS TO THE QUALITY AND PERFORMANCE OF THE PROGRAM IS WITH YOU. SHOULD THE PROGRAM PROVE DEFECTIVE, YOU (AND NOT PRENTICE-HALL, INC. OR ANY AUTHORIZED DISTRIBUTOR) ASSUME THE ENTIRE COST OF ALL NECESSARY SERVICING, REPAIR, OR CORRECTION.

SOME STATES DO NOT ALLOW THE EXCLUSION OF IMPLIED WARRANTIES, SO THE ABOVE EXCLUSION MAY NOT APPLY TO YOU. THIS WARRANTY GIVES YOU SPECIFIC LEGAL RIGHTS AND YOU MAY ALSO HAVE OTHER RIGHTS THAT VARY FROM STATE TO STATE.

Prentice-Hall, Inc. does not warrant that the functions contained in the program will meet your requirements or that the operation of the program will be uninterrupted or error free.

However, Prentice-Hall, Inc., warrants the diskette(s) on which the program is furnished to be free from defects in materials and workmanship under normal use for a period of ninety (90) days from the date of delivery to you as evidenced by a copy of your receipt.

LIMITATIONS OF REMEDIES

Prentice-Hall's entire liability and your exclusive remedy shall be:

1. the replacement of any diskette not meeting Prentice-Hall's "Limited Warranty" and that is returned to Prentice-Hall with a copy of your purchase order, or
2. if Prentice-Hall is unable to deliver a replacement diskette or cassette that is free of defects in materials or workmanship, you may terminate this Agreement by returning the program.

IN NO EVENT WILL PRENTICE-HALL BE LIABLE TO YOU FOR ANY DAMAGES, INCLUDING ANY LOST PROFITS, LOST SAVINGS, OR OTHER INCIDENTAL OR CONSEQUENTIAL DAMAGES ARISING OUT OF THE USE OR INABILITY TO USE SUCH PROGRAM EVEN IF PRENTICE-HALL OR AN AUTHORIZED DISTRIBUTOR HAS BEEN ADVISED OF THE POSSIBILITY OF SUCH DAMAGES, OR FOR ANY CLAIM BY ANY OTHER PARTY.

SOME STATES DO NOT ALLOW THE LIMITATION OR EXCLUSION OF LIABILITY FOR INCIDENTAL OR CONSEQUENTIAL DAMAGES, SO THE ABOVE LIMITATION OR EXCLUSION MAY NOT APPLY TO YOU.

GENERAL

You may not sublicense, assign, or transfer the license or the program except as expressly provided in this Agreement. Any attempt otherwise to sublicense, assign, or transfer any of the rights, duties, or obligations hereunder is void.

This Agreement will be governed by the laws of the State of New York.

Should you have any questions concerning this Agreement, you may contact Prentice-Hall, Inc., by writing to:

Prentice-Hall, Inc.
College Division
Englewood Cliffs, N.J. 07632

Should you have any questions concerning technical support you may write to:

David Meredith
Department of Mathematics
San Francisco State University
1600 Holloway Avenue
San Francisco, CA 94132

YOU ACKNOWLEDGE THAT YOU HAVE READ THIS AGREEMENT, UNDERSTAND IT, AND AGREE TO BE BOUND BY ITS TERMS AND CONDITIONS. YOU FURTHER AGREE THAT IT IS THE COMPLETE AND EXCLUSIVE STATEMENT OF THE AGREEMENT BETWEEN US THAT SUPERCEDES ANY PROPOSAL OR PRIOR AGREEMENT, ORAL OR WRITTEN, AND ANY OTHER COMMUNICATIONS BETWEEN US RELATING TO THE SUBJECT MATTER OF THIS AGREEMENT.

ISBN 0-13-117441-X

CALCULUS CALCULATOR

David Meredith

San Francisco State University

PRENTICE HALL, ENGLEWOOD CLIFFS, NEW JERSEY 07632

Editorial/Production Supervision: **Naomi Nishi**
Cover Design: **Ben Santora**
Manufacturing Buyers: **Paula Massenaro** and **Lori Bulwin**

Printed in the United States of America

10 9 8 7 6 5 4 3 2 1

ISBN 0-13-117441-X

Prentice-Hall International (UK) Limited, *London*
Prentice-Hall of Australia Pty. Limited, *Sydney*
Prentice-Hall Canada Inc., *Toronto*
Prentice-Hall Hispanoamericana, S.A., *Mexico*
Prentice-Hall of India Private Limited, *New Delhi*
Prentice-Hall of Japan, Inc., *Tokyo*
Simon & Schuster Asia Pte. Ltd., *Singapore*
Editora Prentice-Hall do Brasil, Ltda., *Rio de Janeiro*

CONTENTS

PREFACE

CC is a simple yet powerful calculating tool for developing, exploring and testing mathematical ideas and models. It gives easy access to the tools of mathematics: transcendental functions, two and three dimensional graphs, equation solving, differentiation, integration, sequences and series, complex numbers and vectors. Its programming methods can be easily learned by the novice, yet are powerful enough to satisfy the experienced programmer. A quick glance at the Table of Contents will give you some idea of the range of capabilities offered by CC.

You should begin by reading Chapter I, which explains how to back up and protect your program disks. Then proceed to Chapter II for a complete introduction to CC. After reading Chapter II you may want to go directly to the examples in Chapter VI to see many different ways that CC can be used. When you want more detailed information about CC's features, read Chapter III, which will tell you everything you could wish to know. If you want to write programs using CC, Chapter IV explains CC's built-in programming language. Finally, after you become more familiar with CC's basic operations, you may want to study the technical discussion of CC's algorithms and logical structure in Chapter V.

Whenever an entry to CC is presented, it will be printed with a special typeface, like this:

```
x = tan(pi/3)
```

All such lines can be copied directly into CC and executed. You can learn to use CC by copying the examples from this manual.

CC runs under DOS 2.0 or later on IBM©-compatible personal computers equipped with any of the standard graphics systems: CGA, Hercules, EGA, or VGA. A minimum of 512K RAM is required.

Any complex program inevitably raises questions from its users. If you send written questions to me with a stamped, self-addressed return envelope, I will answer as many as I can. Please do not telephone; the expected volume of such calls would overwhelm me. I also welcome bug reports and suggestions for improvements. I will reward anyone who first reports an error to me with a corrected copy of CC.

This project has been generously supported and encouraged by more people than I can name. I am grateful to all those who accepted early versions of CC and patiently reported the many deficiencies they found. The Mathematics Department at SFSU has supported the distribution of these early versions. My colleagues Hal Forsey and J. T. Smith at San Francisco State have provided extensive testing and many useful suggestions. Additional valuable suggestions have come from Daniel Kemp at South Dakota State University, George Corliss at Marquette and James McKenna at the State University of New York, Fredonia. The anonymous reviewers provided by Prentice-Hall were very helpful. The final form of CC owes much to the recommendations of all these pioneering users.

Writing a program and manual is a long process, and I owe much to the patience and understanding of my wife, Idell Weydemeyer, who tolerated my periodic demands for more hardware and software and encouraged me to complete this work.

David Meredith
Department of Mathematics
San Francisco State University
San Francisco, CA 94132

CALCULUS CALCULATOR

CALCULUS CALCULATOR

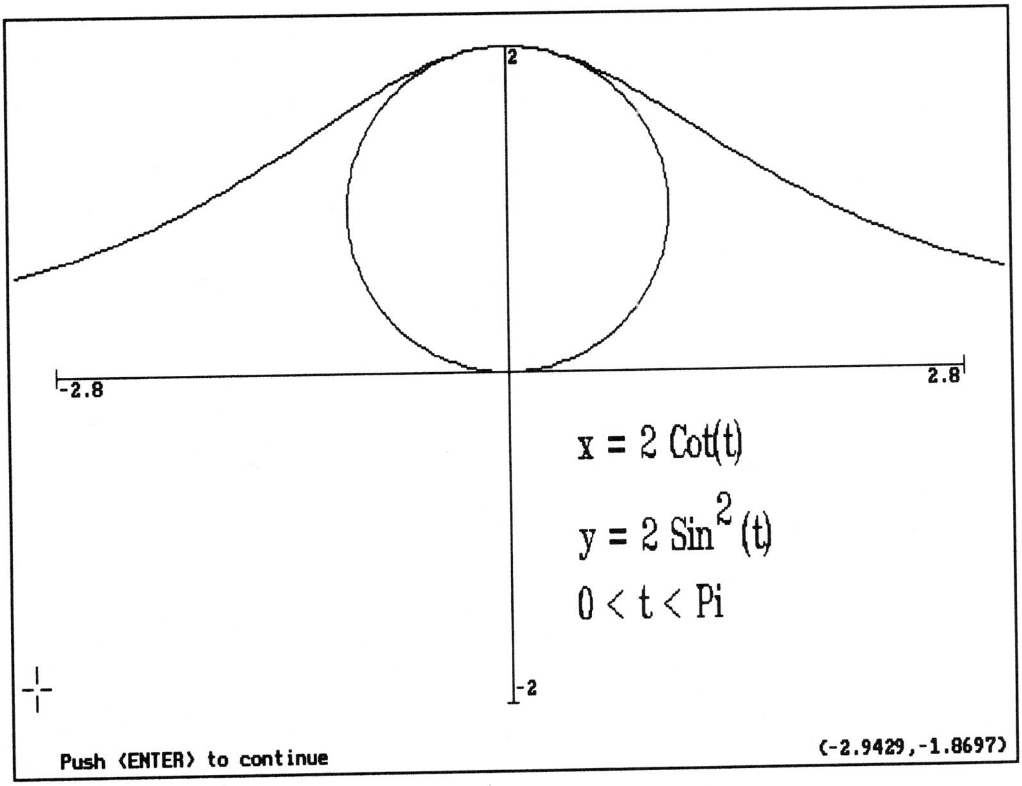

I. BEFORE YOU BEGIN

CC is distributed with this manual and a disk containing three files:

`CC.EXE`	the executable program CC	
`CC.HLP`	a file of help screens for CC	
`DEMO.CC`	an interactive demonstration file	(see II.8)

Before starting to use CC, you should make two copies of the disk, one for backup and one to use. That way, if your working disk goes bad, you will still have the original and one backup disk. If you have a hard disk, you may want to place your working copy of CC there. If you do, you should still make one backup copy on a floppy disk. CC's disk is not copy protected, so you can copy it as many times as you want.

Before you begin copying, you should protect the disk you purchased from accidental erasure. To protect a 5¼" (paper covered) disk, place a write-protect tab or a piece of tape over the small, square notch on the edge of the disk cover. To protect a 3½" (plastic covered) disk, slide the built-in write-protect tab to the position that exposes a small hole in the disk case. Protecting your disk will prevent you from writing any new information onto your original disk or erasing the files already recorded there.

Copying CC to a floppy disk

If you have two disk drives:

Place the original (source) disk in drive `A:` and a blank, formatted disk in drive `B:`.

Type the command `copy A:*.* B:` followed by <ENTER>.

When the computer indicates that the copying operation is complete, remove both disks. All of the files on the source disk from drive `A:` will have been copied to the disk in drive `B:`.

If you have one disk drive:

Place the original (source) disk in the disk drive. Have a blank, formatted disk ready to use.

Type the command `copy A:*.* B:` followed by <ENTER>.

When the computer instructs you to put a disk in drive B:, remove the source disk from the disk drive and insert the blank (target) disk. When the computer instructs you to put the disk into drive `A:`, remove the target disk and replace the source disk in the disk drive. You are using the same disk drive as both drives `A:` and `B:`. The drive acts as drive `A:` when it is reading from the source disk; it acts as drive `B:` when it is writing to the target disk.

When the computer indicates that the copying operation is complete, remove both disks. All of the files on the source disk will have been copied to the target disk.

Copying CC to a hard disk

If desired, make a directory to hold CC's files and change to that directory. Then put the disk to be copied into the disk drive (drive `A:` if there are two) and type the command `copy A:*.* C:` followed by <ENTER>. When the computer indicates that the copying operation is finished, remove the source disk from the disk drive.

Once you have copied your disk, store the original and the backup copy in a safe place.

II. INTRODUCTION TO CC

With CC you can use your computer to calculate mathematical expressions, graph functions, solve equations, and evaluate integrals and derivatives. This chapter will teach you how to use CC to perform these computations. The following chapters explain how to use CC to perform even more complex mathematical operations.

II.1 Starting CC

Throughout this manual, when you read a phrase like "enter the command" or "type the formula," then you must always complete your entry by pressing `<ENTER>`*, which is the key marked* "Enter" *or* "Return" *or* "<—┘". *On the other hand, when you read "press the key", you need not follow this action with* `<ENTER>`*. Just pressing the key will cause CC to perform the indicated operation. From now on we shall omit the explicit instruction to use* `<ENTER>`*, assuming that you will remember that you must terminate all entries except single-key commands with* `<ENTER>`.

Floppy disk system

To start running the program CC from a floppy disk, put your working disk into drive `A:`. If you have only one disk drive, that's the one to use. Then type the commands (each followed by `<ENTER>`):

```
A:
CC
```

Your screen will look like Figure II.1, and you are ready to start calculating. If you are going to use graphs, you should remove your program disk and place a blank, formatted disk in drive A:. See pages 81-82 for more information on this point.

Hard disk system

To start running the program CC from your hard disk, change to the directory holding the files `CC.EXE` and `CC.HLP`, then enter the command:

```
CC
```

followed by <ENTER>. Your screen will look like Figure II.1, and you are ready to start calculating.

When you start CC, you may see a message warning of dangerously low or even insufficient memory. CC may not even load, in which case your screen will display the message "Program too big to fit in memory" or "Runtime Error 203 at xxxx:xxxx." CC is a large program that requires a minimum of 512K RAM memory to operate. It may not run in conjunction with resident programs like disk or print buffers, SideKick©, Windows©, etc. that reduce your available memory below 512K. More information on solving memory problems with CC can be found in section III.12.6.

Take a moment to examine the CC screen (Fig. II.1). At the top is a list of command keys for CC, including the <ESC> key, which is used to quit CC, and the

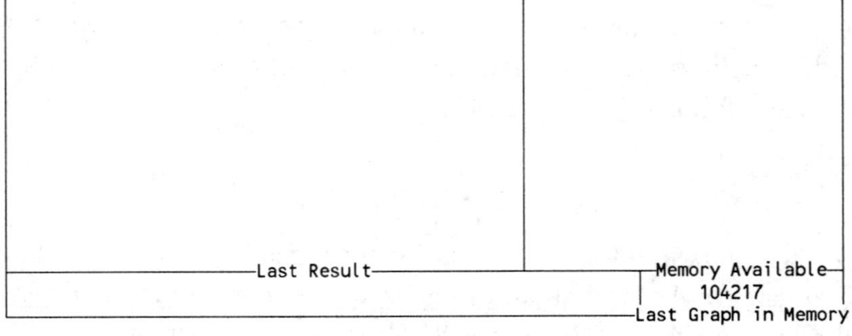

Fig. II.1: CC's Opening Screen

F1	Help key	
F2	Print formulas or graphics	(III.12.1)
F3	Insert a blank line	(III.1)
Shift-F4	Delete a line	(III.1)
F5	Scroll output up	(III.3)
F6	Scroll output down	(III.3)
F7	Save formulas and graphs on disk	(III.12.3)
F8	Load formulas and graphs from disk	(III.12.3)
F9	View last graph	(III.4)
F10	Switch Input/Subroutine Windows	(IV)
Shift-F10	Clear the calculator	(III.1)
<ESC>	Quit the calculator	

CC Command Keys

function key `F1`, which calls up the on-line help screens. The command keys are listed in the box above together with references to an explanation for each one. Anytime you need to be reminded how to use one of CC's operations, try the help key `F1`.

Listed with the function-key reminders on the main screen are the words `Radian` and `Insert`. `Radian` indicates that CC will evaluate trigonometric functions in radian mode. To change to degree mode, enter the command: `Degree`. To return to radian mode, enter the command: `Radian.`

The indicator `Insert` indicates that CC starts in insert mode. If you move the cursor to the middle of a line and start typing, the right half of the line will move over as you type. If you push the key marked `<INSERT>` or `<INS>,` the indicator will change to "Overwrite," and anything you type will write over whatever is already on the screen. Push `<INSERT>` again to return to insert mode.

Below the reminder window are two large rectangles labeled `Input` and `Output`. Your commands will be typed into the Input Window, and your results will appear in the Output Window. Below these two windows is an area labeled `Last Result`. The result of your last command will appear there. To the right of the Last Result Window is the window labeled `Memory Available` which displays the amount of memory (in bytes on the heap) unused by CC. When this number falls below

10,000, you may find that complex operations cannot be performed. When it falls below 5,000, CC will warn you that you are running short of memory.

At the bottom of the Memory Window is a note indicating where your last graph is saved. You can switch from saving the graph in memory to saving it on disk by pressing `Shift-F9`. Depending on which graphics system you are using, this step will increase your available memory between 30,000 and 150,000 bytes. See III.12.6 for more information on memory utilization.

II.2 Running the Demonstration File

Your distribution disk includes a prepared file of formulas demonstrating some of CC's capabilities. This file, `DEMO.CC`, will give you a guided tour of CC.

To run the demonstration file from a *floppy disk*, first start CC, then put a disk with the demonstration file in drive `A:`. (You are using your working disk, and not the original, aren't you?) Press function key `F8`, select the option `(C)alculator Entries` by pressing `"C"`, and enter the file name `A:DEMO.CC`, which will tell CC not only the name of the file but the location of the disk drive where it can be found. The file will be loaded, and you can follow the instructions printed on the screen. Remember to put a blank, formatted disk in drive A: before continuing.

To run the demonstration file from your *hard disk*, first start CC, then press function key F8, select the option (C)alculator Entries by pressing "C", and enter the file name `DEMO.CC`. The file will be loaded, and you can follow the instructions printed on the screen. You do not need a blank floppy disk if you are using a hard disk.

In this file you will see the symbol `//` used in some lines. This is the symbol for a "comment" or "remark." CC ignores anything following `//` in a formula or command line. Comments are used to explain your formulas to other users or to remind yourself why you entered the commands that you used.

For more information about saving and loading files of formulas from CC, see III.12.3. For information about printing calculations and graphs, see III.12.1.

II.3 Calculating with Numbers and Variables

To perform a numerical calculation with CC, just type in the formula you wish to calculate and press <ENTER>. The result of your calculation will appear in two places: after the variable ANS in the Output Window and in the Last Result Window.

Typing is easy with CC, because you can use the cursor (arrow) keys to reposition the blinking cursor in the formula you are typing to make corrections. The <←> (Backspace) and <DEL> keys can be used to erase parts of your formula, or, if you want to retype an entire formula, you can switch to Overwrite mode by pressing the <INS> key and type over your previous work. Pressing <INS> again returns you to insert mode. Your formula is executed when you press <ENTER>.

You can also use the cursor keys to return the cursor to a formula you have already used, change it if desired, and then execute it by pressing <ENTER>.

Try entering and executing the formula:

```
2 + 2
```

The result appears in the Output Window. Now use the up-arrow key to return the cursor to the formula, move the cursor to +, press the <DEL> key, type /, and press <ENTER>. You have changed your formula to `2 / 2` and executed it. Here are some more formulas to try:

`2.3 + 3.6*1.8`	* means multiply
`sin(1.2)`	That is 1.2 *radians*
`ln(0.64)^2`	^ means exponentiation, so ^2 means "squared"

Here are some points to remember.

- Multiplication is denoted by * . Multiplication can also be denoted by a space or, where no ambiguity is possible, by placing two symbols side by side. The following formulas are all equivalent:

```
3.6*1.8    3.6 1.8    3.6(1.8)
   (3.6)1.8    (3.6)(1.8)
```

It is best in the beginning to use * to denote multiplication. Your formulas will be easy to read and unambiguous. Throughout this chapter, multiplication will always be denoted by *, although in later chapters we will drop this symbol so that the formulas look more like traditional mathematics.

- Trigonometric functions are calculated in radian mode unless you specify degree mode. See II.1 above.
- Function names like `Sin` can be entered with upper or lowercase letters. Capital letters are the same as lowercase for CC, so `Sin`, `sin` and even `siN` are all the same.
- Numbers can be entered into CC in either ordinary or scientific notation. The exponent in scientific notation follows the letter `E`. All of the following numbers are equal for CC:

```
3.1    3.10    0.31E1    .031E2    31E-1   310E-2
```

Numbers cannot include blanks. The entry *1.2 E3* will be treated as the product of `1.2` times the variable `E3` (see below for variables).

CC knows the constants π = 3.14159..., e = 2.718... and $i = \sqrt{-1}$, but you must use special keys to enter these values.

Value	Keys	CC symbol
π	alt-p or "PI"	π
e	alt-e	ê
i	alt-i	î

For example, to enter π, hold down the <Alt>-key and press <p>. The box below shows some more CC formulas with their conventional mathematical equivalents.

CC recognizes all the usual mathematical functions like `Cos`, `Ln` and `Atan`. A complete list of CC's predefined functions can be found in section III.2.

CC can calculate with complex numbers too. Remember, i is entered as `alt-i` and displayed on your screen as `î`. Here are some sample complex numbers:

Math	*CC*
$\frac{3.2+4.6}{5.8-8.6}$	(3.2+4.6)/(5.8-8.6)
$\sin^2(\frac{3\pi}{4})$	sin(3*pi/4)^2 *or* sin(3π/4)^2
$\sqrt{3.4}$	sqrt(3.4)
$e^{-2.1}$	ê^(-2.1) *or* exp(-2.1)

Sample CC Formulas

```
î    3 + 4î     2.4 + 3.5î
```

and some formulas using them:

```
1/(1+î)      exp(πî)     î ^ î
```

Almost all of CC's functions accept complex arguments.

Variables

Numerical results can be saved by storing them in *variables*. To assign the value 3 to the variable x, enter:

```
x = 3
```

When you do this, you will see the variable x appear with its value in the Output Window. The number of variables you can define is limited only by your computer's memory. If you have defined more than can be shown in the Output Window at one time, use the function keys F5 and F6 to scroll the output through the window to bring any variable into view.

Variable names can be up to 12 characters long, but they cannot duplicate the name of CC's predefined operators or any previously defined function (see II.3 below). They must start with a letter.

Once defined, variables can be used in calculations. For example, having defined x above, you could now define:

```
y = cos(x)
```

Calculated values that are not explicitly assigned to variables are, by default, assigned to the variable `ANS`. You can use `ANS` in a calculation. If you have just calculated one value and want to store its logarithm in the variable `t`, you can enter:

```
t = ln(ANS)
```

Sometimes you will make a calculation, then remember that you want to save it. Rather than rewriting the line and assigning the formula to a variable, just enter the line:

```
Save = ANS
```

and put the previous result into the variable `Save`.

II.4 Graphing

To graph an expression like x^2 - 1, enter the command:

```
Graph(x^2-1,x)
```

(Fig. II.2). You must specify the variable for graphing, x in this case, to distinguish it from any other variables that may be in your expression. This graph is shown in Figure II.2.

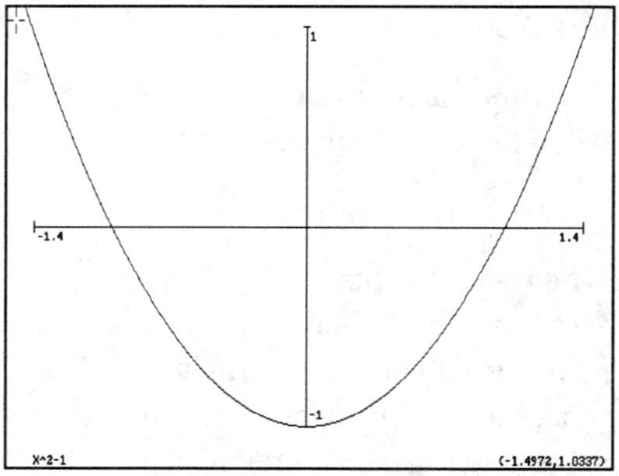

Fig. II.2: y = x^2 - 1

Displayed in the upper-left corner of the graph is a set of crosshairs. The coordinates of the crosshairs are printed at the lower-right corner of the screen. You can use the crosshairs to find the coordinates of any point on the screen by moving them to the desired point with the cursor (arrow) keys. Finer movements can be effected with the shifted cursor keys (or, for those with extended keyboards, with the keys 4 (left), 6 (right), 8 (up) and 2 (down)). Although the crosshairs do not give absolutely accurate results (their accuracy is limited by the resolution of your graphics screen), they are a valuable tool for estimating points where graphs intersect or have maxima or minima.

If you try to graph the function $2x^2 - 2$:

```
Graph(2*x^2-2,x)
```

you will notice that the graph does not fit on the screen (figure II.3).

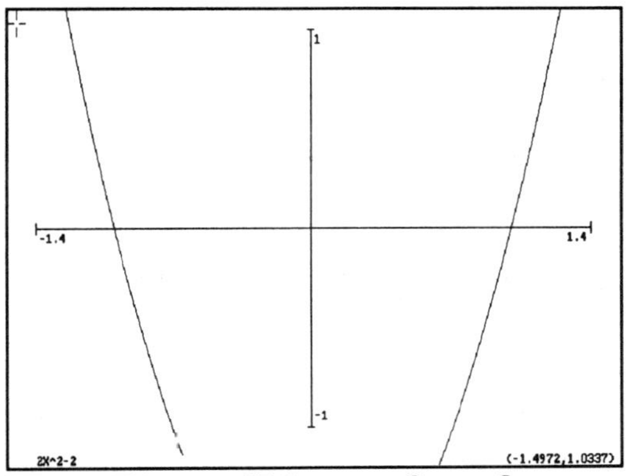

Figure II.3: $y = 2x^2 - 2$

When CC starts, the horizontal or x-axis on the graphics screen extends from -1.4 to 1.4 and the vertical or y-axis from -1 to 1. To graph `2*x^2-2` on this horizontal axis, you need a vertical axis with limits -2 and 2. To create one, use the command:

```
Window(-1.4,1.4,-2,2)
```

The Window command specifies the limits of the graphing screen. If you try the command `Graph(2*x^2-2,x)` again, you will see more of the parabola (figure II.4). CC can also graph polar and parametric equations; place labels on graphs; plot points, line segments and rectangles; and fill bounded regions. For more details on graphing, see III.4.

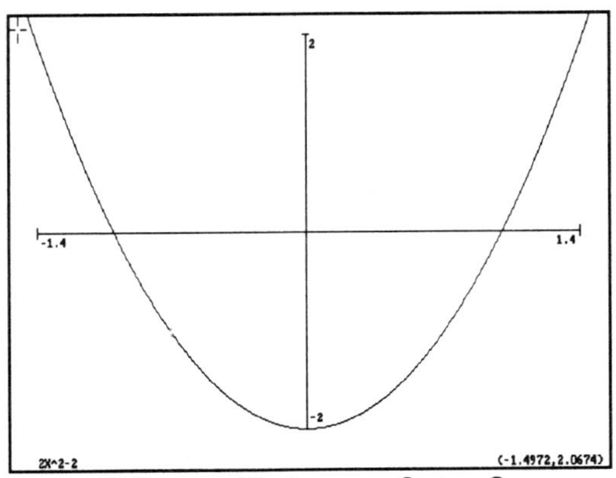

Figure II.4: $y = 2x^2 - 2$

II.5 Differentiation

To find the derivative of an expression like t^2 - 1 at the point t = 2:

$$\left.\frac{d}{dt}(t^2-1)\right|_{t=2}$$

enter:

```
dif(t^2-1,t=2)
```

The symbolic derivative can also be computed by CC. To find the symbolic derivative of t^2 - 1, enter the command:

```
dif(t^2-1,t)
```

(This only works if you have not already given t a value. For more information on differentiation, see III.5.) CC differentiates a function at a point by first finding its symbolic derivative and then evaluating the derivative at the desired point. This process is as accurate as can be; CC's derivatives are as accurate as CC's functions.

II.6 Integration

CC can integrate any expression you can define. To evaluate the integral:

$$\int_1^2 (t^2-1)dt$$

enter the formula:

```
In(t^2-1,t = 1 to 2)
```

CC uses numerical approximations to compute integrals. Thus your answer may not be perfectly accurate, although you can expect at least six digits accuracy for most integrals. If you compute the same integral twice, you may receive slightly different answers, because CC's integration algorithm (V.3.1) incorporates a stochastic element to avoid errors on periodic functions. Even so, the difference between two calculations will be less than the six digit accuracy of CC's integration algorithm. For more information about controlling the accuracy of the integration operator, see III.12.5.

II.7 Solving Equations

CC will solve almost any equation you can think of. Just type in the command `Solve` followed by the equation and a guess at the solution, and CC will compute a very accurate estimate of the solution. For example, to solve the equation:

$$x^3 + 3 = x^2 + 1$$

using $x = 1$ as your first guess, you would enter the formula:

```
Solve(x^2+3 = x^3+1, x = 1)
```

The solution 1.695621 is returned as the value of the variable x in the Output Window.

Before solving an equation, it is sometimes useful to graph it first. Consider the equation $\cos(x) = x$. This equation has a single solution between $x = 0$ and $x = 1$. To see that this is so, use the following commands to graph the functions x and $\cos(x)$ and observe that they intersect once above the interval (0, 1):

```
Window(-1,2,-1,2)
Graph(x,x)
Graph(cos(x),x)
```

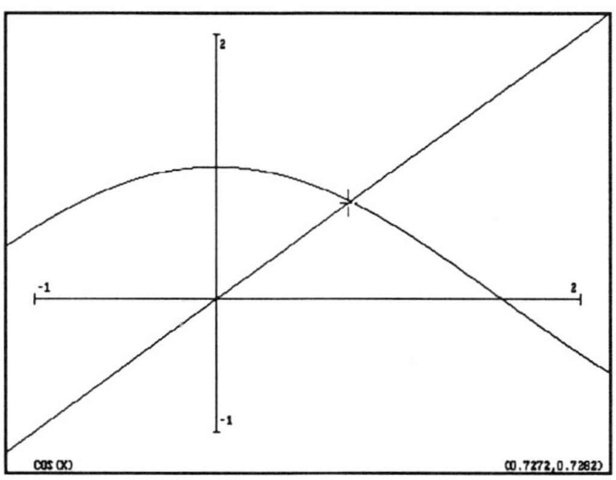

Figure II.5: Intersection of $y = x$ and $y = \text{Cos}(x)$ at $x \approx 0.73$

(Figure II.5). The x-coordinate of the

intersection point is the solution to the equation $\cos(x) = x$. Use the crosshairs to estimate the x-coordinate of the intersection point. This approximation will be about 0.73 (Figure II.5). Using this value as the initial approximation, CC will solve the equation if you enter:

```
Solve(cos(x) = x, x = 0.73)
```

The solution 0.739085 is stored in the variable x and displayed in the Output Window. You might experiment with values for the initial approximation other than 0.73 to see how far away from the final answer CC can start and still solve the equation. Some equations require a more accurate initial estimate than others for successful solving, so it is always safest to give CC the best initial approximation that you can find.

For more information on solving equations, including directions for solving systems of equations and defining implicit functions, see III.7. To learn how to control the speed and accuracy of equation solving, see III.12.5.

II.8 Three Dimensional Graphs of Surfaces

To graph the surface defined by the equation $z = f(x, y)$, you must first decide what region you want to display. Let's suppose you wanted to graph the function:

$$f(x,y) = x^2 - y^2$$

over the region $-1 \le x \le 1$, $-1 \le y \le 1$. You must also decide how much of the vertical range of the graph you wish to show. On the specified region, the values of $f(x,y)$ are between -1 and 1, so you may want to use a vertical range of (-1,1). Finally, you must decide how many values of the function to plot in the x and y directions. The smallest useful value is usually 10; the largest permitted is 40. We will use 15 in this

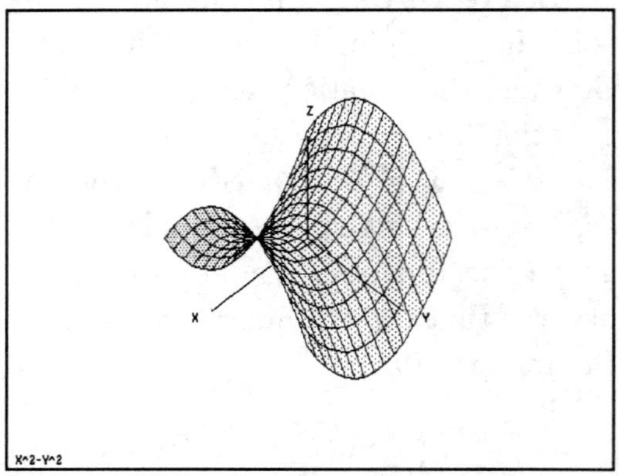

Figure II.6: $f(x,y) = x^2 - y^2$

example. The more values you plot, the smoother the resulting graph and the longer it takes to create.

All of these parameters are used in the three-dimensional graphing command:

```
Graph3d(x^2- y^2; x = -1 to 1; y = -1 to 1; -1 to 1 step 15)
```

With this command, you are telling CC to use values for x and y between -1 and 1, to graph values between -1 and 1, and to use 15 subdivisions. The vertical values do not require a variable name. Figure II.6 shows the resulting graph.

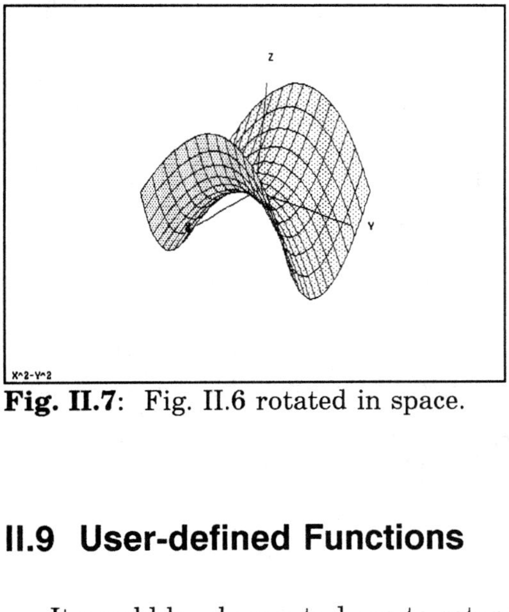

Fig. II.7: Fig. II.6 rotated in space.

Once you have a three dimensional graph on your screen, you can rotate it six ways by pressing the keys: x, X, y, Y, z and Z. Try it. Figure II.7 shows Figure II.6 after pressing the keys `Y Y Y z z`.

The graphing command above uses a new character, the semicolon. It is used purely for purposes of clarity and could be replaced by a comma. See III.1 for more information on punctuation in CC.

II.9 User-defined Functions

It would be clumsy to have to retype a long expression every time you wanted to use it. CC allows you to define your own functions, which can then be used just like its built-in functions. Once defined, functions can be graphed, differentiated, integrated, solved, and used to define other functions. To define a function, just enter it in usual mathematical notation, for example:

```
f(x) = x^2 + 1
```

The function `f(x)` is called a *user-defined function*, to distinguish it from the built-in functions like `Sin` and `Sqrt`, and to distinguish it from the subroutine functions that will be used for programming.

You will not see the variable x in the Output Window after you make this definition, because x doesn't have a value. It is a *bound variable*, a placeholder in the function definition. You can use the same symbol for the bound variable in several different functions without confusing CC. You can even use the same symbol for a free variable, and CC can tell which you mean from the context.

You evaluate $f(x)$ by substituting a value for x. For example, you could compute:

```
a = f(5.4)
```

Once you have defined a function, you can use it exactly like one of CC's built-in functions. Here are some sample expressions using $f(x)$:

`b = 2*f(3.4-2.1)^2`	Expression using $f(x)$
`c = dif(f(x),x=2)`	Derivative of $f(x)$ at $x = 2$
`k = in(f(x),x = 0 to 3)`	Integral of $f(x)$ from 0 to 3
`g(x) = dif(f(x),x)`	Symbolic derivative of $f(x)$

The first three lines above produce values; the fourth defines a new function that is the derivative of $f(x)$.

Function names can be up to 12 characters long, but they cannot duplicate the name of any of CC's built-in functions and operators or any previously defined variable. They must begin with a letter.

One function can be defined in terms of another, and function definitions can use variables too. You could define:

```
g(x) = f(a*x) + 1
```

Note that the value of this function depends on the value of a when the function is evaluated, not when it is defined. For more information about functions, see III.11.

III. CALCULATING WITH CC

This chapter explains all of the commands, functions and operators that can be entered through CC's Input Window. There are sections on graphing functions; integrating and differentiating; solving equations and systems of equations; defining sums and products; and using lists for vectors and statistics. The chapter concludes with a discussion of loading, saving and printing data with CC, controlling the speed and precision of integration and equation solving, customizing CC's configuration and memory management.

III.1 Entering Formulas into the Input Window

To enter a formula into the Input Window, just type it and press <ENTER>. You can use the cursor keys, the <BACKSPACE> key and the <DEL> key to correct your formula (see II.2). The <INSERT> key toggles the Insert/Overwrite mode; the default is Insert mode. Once you have entered a formula into the Input Window, you can use it again by moving the cursor to it. Then you can modify the formula if you wish and re-execute it by pressing <ENTER>.

Entries can be created that are longer than the width of the input window. Just keep typing, and the entry will spill over onto succeeding lines, indented one space. Each entry can be as long as 255 characters. *Do not push* <ENTER> *while entering a formula.* <ENTER> terminates the creation of a formula and causes it to be executed.

To open up free lines in the list of formulas, press function key F3. To erase a line, press Shift-F4. To erase all the lines you have entered (including the subroutines--see Chapter IV below), press Shift-F10 and respond "Y" when asked to confirm that you wish to clear all entries. You can add comments to a line in CC, or even enter a line which is entirely a comment, by preceding the comment part with two slashes `//`. For example:

```
Tan(3.4)    // Slope of the tangent line
```

Comments make your calculations more understandable, whether you are printing your calculations or just want to jot some notes to yourself while you are working. Comments can also be used to give directions when you are preparing a worksheet or lesson for someone else to complete.

For purposes of clarity within long formulas, CC permits a variety of separating punctuation marks. The symbols ","(comma), ";"(semicolon), "=" (equals) and the words "to" and "step" are equivalent for CC. You can use any of them to make your formulas easier to read. The following two formulas are equivalent:

```
Sum(n^2; n = 1 to 21 step 2)
Sum(n^2,n,1,21,2)
```

For Pascal programmers, semicolons at the ends of lines are accepted and ignored.

III.2 Mathematical Calculations

CC is an environment for mathematical calculating, and the format of its formulas is as close to ordinary mathematical notation as can be with the limitations of writing on a single line. If you are not sure how to enter a formula, just write it as you normally would in mathematics. It will probably compute successfully. In technical terms, CC's calculating language is highly orthogonal, meaning that any expression which ought to make sense in a context can be used there. There are few limitations on complexity. Wherever a number would fit, you can place a numerically-valued expression. Any function definition can use previously defined functions and variables. Even undefined functions and variables can be used, and they are retained literally as typed until they are defined.

III.2.1 Numbers and Arithmetic

CC accepts integer and real numbers in most conventional formats. You can enter `2, 20, 0.2, .2, -2, -0.2, -.2,` etc. You can also enter numbers in scientific notation: `2E5`, `.02E-5`. The only limitation is that blanks cannot be included in numbers. See II.2 for details.

The five basic arithmetic operations are denoted:

+	Addition
-	Subtraction
*	Multiplication
/	Division
^	Exponentiation

For example, to compute $2(3 + 4)^2$, enter `2*(3 + 4)^2`. CC follows the usual algebraic conventions for order of operations. The order specified by parentheses is always followed, but when a formula is ambiguous exponentiation precedes multiplication and division, which precede addition and subtraction. Operations with the same precedence level are performed left to right, except for exponentiation, which is performed right to left. Exponentiation can be performed with any real or complex exponent, but internally CC uses fast multiplication for positive integer exponents and logarithms for other exponents.

The multiplication symbol can be replaced by a blank or even omitted when it is clear from context. The multiplication symbol is necessary only in the following contexts:

- between two numbers:
  `34` means thirty-four, `3*4` and `3 4` equal `12`;
- between two variable names:
  `xy` is a single variable, `x*y` and `x y` are products of two variables;
- between a variable name and a number:
  `x3` is a single variable, `x*3` and `x 3` are products (but `3x` is a product because a variable name cannot start with a number);
- between a variable name and a left parenthesis:
  `f(3+4)` is a function, `f*(3+4)` and `f (3+4)` are products;

When in doubt, put in the multiplication symbol. Your formulas may look a bit less like traditional mathematics, but they will be easier to read and potential ambiguities will be avoided. *The most common error in CC is to forget the multiplication* *symbol between two variables.* It is easy to write xy when you mean x*y, but CC will treat xy as a single, two-character variable name.

Formulas which are undefined or infinite, like 1/0 or Tan(pi/2), are given the value UNDEFINED. UNDEFINED is a number in CC, albeit with some strange properties. All formulas (except x==a and x<>a--see IV.3.1) including an undefined value return the result UNDEFINED. You can use UNDEFINED in your own formulas, which might be useful in a subroutine to test if the outcome of a calculation was defined. Try entering 1/0 to see an undefined result. Can you find any other undefined formulas?

In addition to UNDEFINED, CC includes the built-in constants π, e, TRUE and FALSE. TRUE and FALSE will be discussed in Chapter IV; π equals 3.141592535897932385 and e equals 2.718281828459. The number π can be entered as Alt-p or PI, and e must be entered as Alt-e. It appears on the screen as ê.

For those who need them, CC can calculate with complex numbers as well as real numbers. The imaginary number $i = \sqrt{-1}$ is entered as Alt-i and appears on the screen as î. Complex values are entered just as you would normally write them: 2.3 - 4.2î for example. Complex numbers can be used in all operations and functions where a complex argument would make sense. You can even integrate with complex limits (the integral is computed on a straight line segment in the complex plane between the limits), and you can solve equations for complex roots. The examples below illustrate complex formulas in CC.

Formula	*Value*	
(2 +3î)/(4 - 5î)	0.170732+0.536585î	
(-2)^(0.5)	1.414214î	
ê^(πî)	-1-7.27596E-12î	approximately -1
î^î	0.20788	

The symbols ê and î are used for special values instead of the ordinary letters e and i, because using the ordinary letters would make them reserved symbols which then could not be used as variable names.

III.2.2 Algebraic functions

CC has the following algebraic functions operating on real and complex values:

Algebraic Functions

Function	*Meaning*	*Example*	*Result*
²	square	`3²`	`9`
`Sqrt`	square root	`Sqrt(5)`	`2.236`

Note: ² is entered as Alt-2.

Computing Real Odd Roots of Negative Numbers

Computing odd roots of negative numbers is a special problem for CC, since it treats real numbers as a special case of complex numbers. If x is a positive number and n a positive integer, then `x^(1/n)` returns the positive n'th root of x. However, if x is negative, then this expression may not return the answer you want. In particular, if n is odd and x negative, then `x^(1/n)` returns the primitive complex n'th root of x, not the real n'th root. Try `(-2)^(1/3)`: CC returns 0.629961 + 1.091124î. Of course, if you entered `-2^(1/3)`, CC would return -1.25992, the real cube root of -2, because exponentiation precedes negation.

You can construct a function that returns the real n'th root of any non-zero real number x, provided one exists:

```
Root(x,n) = x/abs(x)*abs(x)^(1/n)
```

For example, to graph the cube root of x, enter the command:

```
Graph(Root(x,3),x)
```

III.2.3 Transcendental Functions

CC has all standard transcendental functions, each of which can be used on real and complex numbers. The branches of the complex functions have been selected so that the inverse functions work correctly when possible.

Trig and inverse trig functions

Function	*Meaning*	*Example*	*Result*
`Sin`	sine	`Sin(pi/4)`	`0.707`
`Cos`	cosine	`Cos(0)`	`1`
`Tan`	tangent	`Tan(pi/4)`	`1`
`Cot`	cotangent	`Cot(pi)`	`UNDEFINED`
`Sec`	secant	`Sec(2*pi/3)`	`2`
`Csc`	cosecant	`Csc(3.26)`	`-8.465`
`ATan`	inverse tangent	`ATan(1)`	`0.785`
`ASin`	inverse sine	`ASin(1/Sqrt(2))`	`0.785`
`ACos`	inverse cosine	`ACos(1)`	`0`
`ATan2`	polar angle	`ATan2(3,-4)`	`-0.644`

Hyperbolic and inverse hyperbolic functions

Function	*Meaning*	*Example*	*Result*
`Sinh`	hyperbolic sine	`Sinh(1.2)`	`1.509`
`Cosh`	hyperbolic cosine	`Cosh(-0.3)`	`1.045`
`Tanh`	hyperbolic tangent	`Tanh(2)`	`0.964`
`ASinh`	inverse hyper. sine	`ASinh(-2.1)`	`1.487`
`ACosh`	inverse hyper. cosine	`ACosh(-2.1)`	`1.4+3.1î`
`ATanh`	inverse hyper. tangent	`ATanh(2)`	`0.54-1.6î`

Exponential and logarithmic functions

Function	*Meaning*	*Example*	*Result*
`Exp`	exponential	`Exp(2.3)`	`9.974`
`Ln`	natural logarithm	`Ln(1/2)`	`-0.693`

Notes:

- Trigonometric functions of a real variable assume that the argument is in radians or degrees, whichever mode is displayed with the function-key reminders near the top of the screen. You can change the mode by entering the command `Degree` or `Radian` through the Input Window. When the argument is complex or will return a complex result, CC always computes in radians.
- The function `ATan2(x,y)`, the polar angle of the vector (x, y), returns a value in the range $(-\pi, \pi]$ if x and y are real and at least one is non-zero. Otherwise the value is `UNDEFINED`.
- The expression `exp(3)` means e^3, which can alternately be entered into CC as `ê^3`. Generally it is faster and more accurate to use the function `exp(3)` than to use the exponential `ê^3`.

III.2.4 Miscellaneous Functions

The functions listed here all take numerical arguments. In addition, CC has functions which take lists for arguments (see III.10) and operators like integration that take functions for arguments (see III.4-III.7).

Miscellaneous Functions

Function	*Meaning*	*Example*	*Result*
`Abs`	absolute value	`Abs(-3)`	`3`
`Re`	real part	`Re(4-3î)`	`4`
`Im`	imaginary part	`Im(4-3î)`	`-3`
`Int`	integer part	`Int(3.4)`	`3`
`Frac`	fractional part	`Frac(3.4)`	`0.4`
`Div`	integer division	`18 Div 4`	`4`
`Mod`	remainder	`18 Mod 4`	`2`
`Par`	parity = $(-1)^n$	`Par(5)`	`-1`
`!`	factorial	`17!`	`3.56E14`
`LnGamma`	Ln(Gamma(.))	`LnGamma(4)`	`1.792`
`Binom`	binomial coefficient	`Binom(5,3)`	`10`
`Random`	random number	`Random`	`0.747`
		`Random(4)`	`3`

Notes:

- If $n \geq 0$, then `Int(n)` is the largest integer less than or equal to n. If $n < 0$, then `Int(n)` equals the smallest integer greater than or equal to n.
- `Frac(n)` is defined by the relation `Int(n)+Frac(n) = n`.
- `Int` works on complex numbers too: `Int(a+bî)=Int(a)+Int(b)î`.
- `a div b` is the integer quotient of a by b: `a div b = Int(a/b)`. `a mod b` is the remainder of a divided by b. `Div` and `mod` work only on positive integers.
- `Par(n)` means $(-1)^n$. It is useful in defining power series (cf. III.9).
- `Random` with no arguments returns a random real in the range (0,1); if n is a positive integer, then `Random(n)` returns a random integer in the range 1..n.

III.2.5 Binary and Hexadecimal Numbers

CC will accept and display integers in binary and hexadecimal format. Binary numbers are preceded with #: for example, `#110001101`; hexadecimal numbers are preceded with `$`: for example, `$3a24`. Numbers up to 32 bits (8 hex digits) may be entered, but binary values longer than 15 bits or digits are displayed in hexadecimal notation. Binary and hexadecimal numbers are positive 32 bit constants, and these numbers can be added, subtracted, multiplied and negated. The results are computed modulo 2^{32}, so `-$2 = $FFFFFFFE`. Integer division with `div` and `mod` work as expected, but real division with `/` returns a decimal result. When binary and hexadecimal values are combined with real numbers or used in transcendental functions, a real result is returned. When integer, binary and hexadecimal values are combined, they return a hexadecimal value if one of the factors is hexadecimal, otherwise they return a binary value if one of the factors is binary.

To transform one of these numbers to another form, use the functions:

Binary and Hexadecimal Numbers

Function	*Meaning*	*Example*	*Value*
`Integer`	n in base 10	`Integer($11)`	`17`
`Hex`	n in base 16	`Hex(21)`	`$15`
`Bin`	n in base 2	`Bin($a)`	`#1010`

To mimic the usual conventions of computer arithmetic, integers are numbers without decimal parts in the range -2^{31} to 2^{31} -1 and hexadecimal and binary numbers begin at 0 and go up to 2^{32}-1. Also, hexadecimal and binary numbers in the range 2^{31} .. 2^{32}-1 are considered equal to integers in the range -2^{31} .. -1. Thus `Hex(-2)` = $FFFFFFFE

Integer, hex and binary numbers can be combined with `and` and `or`, and you can apply the function `not` to them. Note that `not(n)` is the same as $-n$-1. All types of numbers—hexadecimal, binary, integer, real, and complex—can be used together in calculations, with the type of the output determined by the types of the inputs. Here are some samples of binary and hexadecimal operations.

Entry	*Result*
`$aa*$bb`	`$72CE`
`#11011+#10110`	`#110001`
`$abcd or $1234`	`$BBFD`
`$abcd and $1234`	`$204`
`not(#110011)`	`$FFFFFFCC`
`hex(50000)`	`$C350`
`bin(50000)`	`#1100001101010000`
`integer($abcd)`	`43981`

III.3 Variables

CC uses variables to store the results of calculations. To store the value of a formula in a variable, put `<var-name> =` before the expression to be evaluated. Variable names must start with a letter and include only letters and numbers. They can be up to 12 characters long, and they are not case sensitive. Here is an entry that assigns a computed value to the variable `a`:

```
a = 2 + 3
```

Variables and their values are listed in the Output window.

Once defined, variables can be used like constants in expressions. The next two lines define the variable `a` and use it to define another variable `b`:

```
a = 2 + 3
b = a*4
```

The sequence just displayed could be changed slightly to:

```
a = 2 + 3
a = a*4
```

The first line places 5 in *a*; the second line calculates the value 5*4 and places the result 20 in *a*. A new value can always be placed in a variable that is already defined, but doing so erases the previous value. A variable can hold only one value at a time. (But see III.10 for information about lists, which store many numbers in a single variable.) After executing these two lines, the variable *a* will equal 20.

Values in CC are created and stored only by executing lines in the Input Window, and the values depend on what lines have been executed and in what order. It does not matter what lines are written there. In the last example, the first line a = 2+3 does not disappear from the calculator when the second line a=a*4 is executed. But the effect of the first line is overridden by the later execution of the second line.

CC has limited symbolic capabilities. If an undefined variable is used in an expression, then the expression returns a symbolic result (which may be stored in another variable). Suppose that *x* is an undefined variable name. The consequences of using this undefined variable are illustrated by the following formulas.

Formula	*Meaning*
`y = 3 + x`	symbolic result since x undefined
`x = 4`	now x is defined
`y`	returns 7, but y still symbolic
`x = 7`	new value in x
`y`	returns 10, but y still symbolic
`y = y`	store 10 in y

When these lines are executed, *y* first receives the formula $3 + x$ as its value. Until *y* is assigned a new value by appearing on the left of an assignment statement, *y* will retain this value. Then *x* is given a value in the second line, so when *y* is executed in the third line, the value 7 is returned in the variable *ANS*. (Remember, formulas not assigned explicitly to a variable are implicitly assigned to *ANS*.) Similarly, after 7 is stored in *x* and *y* is executed again, 10 is stored in *ANS*. Finally, *y* is executed again, resulting in a value of 10, but this time the value is stored back into *y*. Storing this new value in *y* eliminates the old value of $3 + x$, since a variable can hold only one value. At the end of this sequence, 10 is stored in *y* .

Symbolic values for variables are useful mostly for viewing the definitions of functions and derivatives. (See III.5).

You must be cautious about combining redefinition of variables with undefined variables. If you enter the following lines for a previously undefined variable *a*:

```
a = a+3
a
```

you will get an error message indicating possible infinite recursion. Clear the error message by pressing any key, and you can continue calculating. More information about variables can be found in V.2.2.

If you have more variables than the Output Window will display, you can scroll them by pressing F5 and F6. Shift-F5 and Shift-F6 page through the variables.

To eliminate a variable--to undefine it--use the command:

```
Forget(v)
```

where v is the name of the variable. This command will also eliminate a function definition (cf. III.11 below). The command `Forget` can be applied to more than one variable. For example, to delete the variables a, b and c, enter: `Forget(a,b,c)`.

III.4 Graphing

CC includes extensive facilities for graphing functions in two and three dimensions. Sections III.4.1-III.4.5 explain two-dimensional graphs, including polar and parametric graphs and options for labeling graphs. Three dimensional graphs are discussed in section III.4.6, and sections III.4.7-III.4.8 demonstrate how to save graphs on disk or print them.

Before you graph any function on the plane, you must specify the region in the plane you wish to view with the command:

```
Window(a,b; c,d)
```

The expressions a, b, c and d must be real numbers or real-valued expressions such that $a < b$ and $c < d$. `Window` tells CC that the visible screen corresponds to a portion of the xy-plane with $a \leq x \leq b$ and $c \leq y \leq d$. When CC starts, a default window is in place: $1.4 \leq x \leq 1.4$ and $1 \leq y \leq 1$. To make your circles and squares look correct--that is, to make one vertical unit the same physical distance on your screen as one horizontal unit--select window parameters so that the horizontal axis is 1.4 times as long as the vertical axis.

The window command does not affect the graphs drawn by `Histogram` (III.10.3-.2) or `Graph3d` (III.4.6), but it does control the visible portion of the result from `ListG`, `Mark` and `Plot`.

If you ask CC to graph a point outside the window, it ignores the command and continues without any error message. Thus functions which are undefined on part of the graphing domain can be graphed.

You can resize the graphics window at any time, but doing so erases the screen. You can also erase the screen without resizing the window with the command:

```
Erase
```

If you want to create a graphics screen without axes, then issue the command:

```
AxisOff
```

before drawing the first graph on an empty screen (either before or after executing `Window` or `Erase`). All screens created after you execute `AxisOff` will omit axes.

To start showing axes again, enter the command:

```
AxisOn
```

Until you erase or resize the graphing window, each two dimensional graph overlays the previous ones. (Three dimensional graphs cannot be superimposed.) This allows graphs of functions to be compared. On color screens successive graphs are different colors; on monochrome screens they are drawn with different line styles. See III.4.5 to learn how to stop automatic changing of colors and styles.

Graphs are actually constructed by computing a series of points and connecting them with straight line segments. If the computed points are too close together, then the different line styles used for drawing on monochrome screens may not be distinguishable.

Three general conventions about graphing:

- A graphing process (or any other process in CC) can be interrupted safely at any time by pressing the <ESC> key.
- The last graphics screen constructed can always be recalled by pressing function key F9. See III.12.6 for more information on how this operation affects memory utilization.
- Most two-dimensional graphics screens include crosshairs. You can move these about the screen, using the arrow keys for gross movements and the shifted-arrow keys for fine movements, to find the coordinates of any point on the

screen. If you are using an extended keyboard, then use the keys <2>, <4>, <6> and <8> for the shifted-arrow keys. Diagonal movements can be effected with the keys <Home>, <End>, <PgUp> and <PgDn> and their shifted equivalents <7>, <1>, <9> and <3>. The coordinates of the crosshairs are always displayed in the lower right corner of your screen, so the crosshairs can be used to locate points on your graph like intersections and extrema.

After exiting the graphics screen and returning to the Input Window, you can recover the coordinates of the last position of the crosshairs with the commands:

```
CrossX        CrossY
```

These functions return the value `UNDEFINED` if the last graphics screen was a histogram or a three-dimensional graph or if no graphics screen has been created.

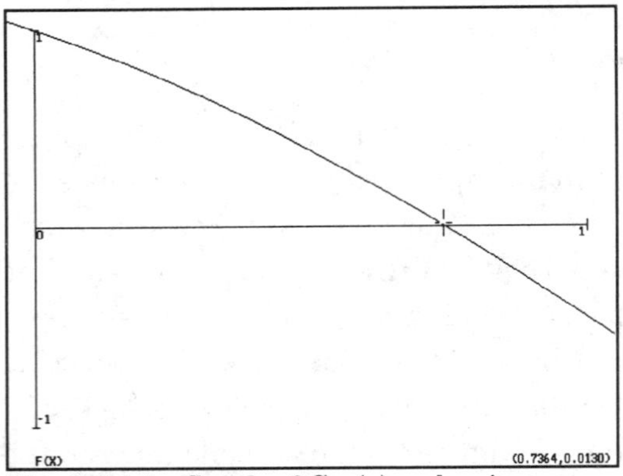

Fig. III.4.1: Graph of Cos(x)-x showing crosshairs positioned at the zero.

Recalling the crosshair coordinates is useful for solving equations. To solve the equation $f(x) = 0$, first graph the equation, move the crosshairs to the point where the graph crosses the x-axis (Fig. III.4.1), press <ENTER> to return to the Input Window, and use `CrossX` as the first guess for the solution to the equation.

```
f(x) = cos(x) - x          // equation to solve
Window(0,1,-1,1)
graph(f(x),x)              // Move crosshairs to root
Solve(f(x) = 0, x = CrossX)
                   // First guess is x-coordinate of crosshairs
```

III.4.1 Cartesian graphs

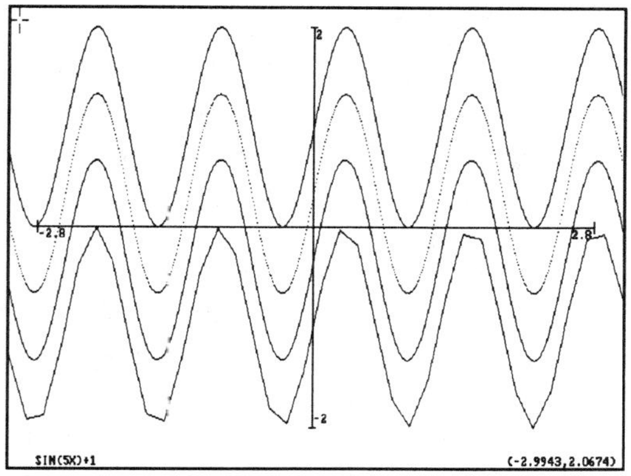

Fig. III.4.2: Curves drawn with Graph, DotGraph, QuickG and Sketch. Examine peaks to distinguish Graph from QuickG.

To graph an expression like sin(5*x*) with respect to *x*, use one of the commands:

```
Graph(sin(5*x)+1,x)
DotGraph(sin(5*x)+1/3,x)
QuickG(sin(5*x)-1/3,x)
Sketch(sin(5*x)-1,x)
```

(Fig. III.4.2). These commands all graph a function *f*(*x*), but in slightly different ways. `Graph` computes a value of *f*(*x*) for *x* corresponding to every pixel on the x-axis and then connects the resulting points (*x*, *f*(*x*)). `DotGraph` computes the same points but just plots them without connecting them with lines. For most graphs there is little difference, but `DotGraph` does not make phony connections across singularities. The command `QuickG` works like `Graph` but only computes values for every third pixel on the *x*-axis. It works three times faster than `Graph`, hence the name. Even faster is `Sketch`, which only computes values for every 20'th pixel. Use `Sketch` to graph a function that is slow to evaluate.

III.4.2 Polar graphs

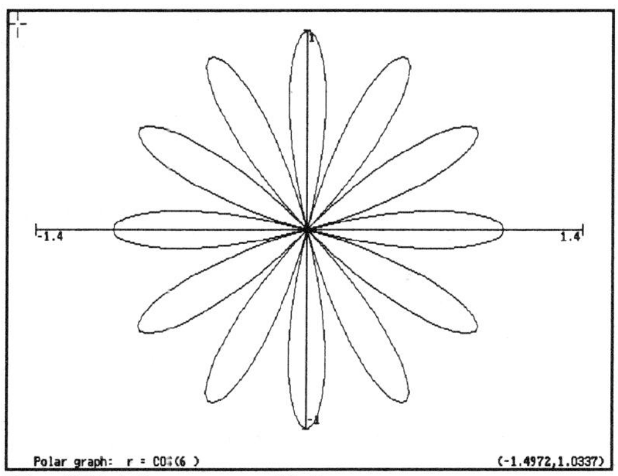

Fig. III.4.3: r = cos(6Θ)

To graph a polar function, you must specify the function to be graphed and the domain of the variable. For example, to graph $r(\Theta) = \cos(6\Theta)$ for the domain $0 \leq \Theta \leq 2\pi$, use the command:

```
PolarG(cos(6Θ),Θ = 0 to 2π)
```

(Fig. III.4.3). Remember, π is entered

with `Alt-p` and Θ with `Alt-t`.

Like other curves, polar graphs are drawn in the existing window. It may be necessary to adjust the window to hold the entire graph, as with this limaçon (Fig. III.4.4):

```
Window(-1,2; -1.5,1.5)
PolarG(0.5+cos(t),t=0 to 2*pi)
```

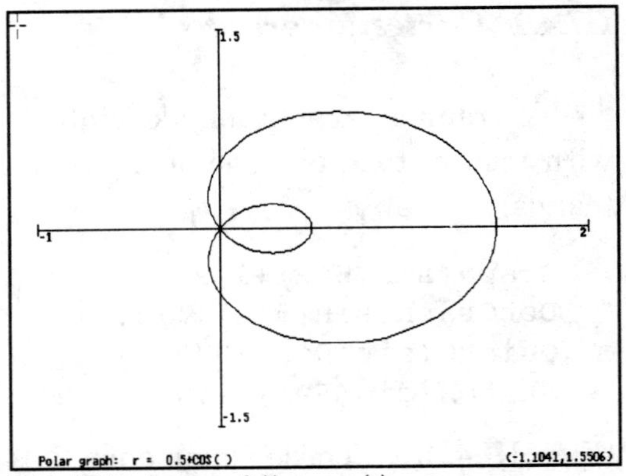

Fig. III.4.4: r = 0.5 + cos(t)

Polar curves are drawn with 200 line segments by default, but you can change this value by adding your own value to the `PolarG` command. Notice how changing the number of line segments changes the graph of even a simple function like r(Θ) = 1 (Fig. III.4.5):

```
Window(-1.4,1.4; -1,1)
PolarG(1, t=0 to 2π, 100)          // circle
PolarG(1, t=0 to 2π, 5)            // pentagon
PolarG(1, t=0 to 4π, 5)            // 5-pointed star
```

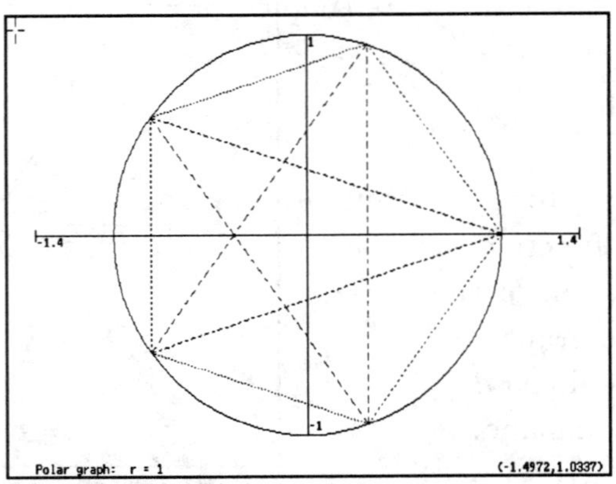

Fig. III.4.5: r(Θ) = 1

III.4.3 Parametric graphs

Parametric curves are graphed almost like polar equations. You must have the correct size window defined, select a domain for the parameter, and optionally choose the number of points to graph. For example, the commands:

```
window(-1.4,1.4; -1,1)
paramg(sin(5*t),sin(6*t);
     t=0 to 2π)
```

graph the Lissajous curve: $x = \sin(5t)$, $y = \sin(6t)$, over the domain $0 \le t \le 2\pi$ (Fig. III.4.6).

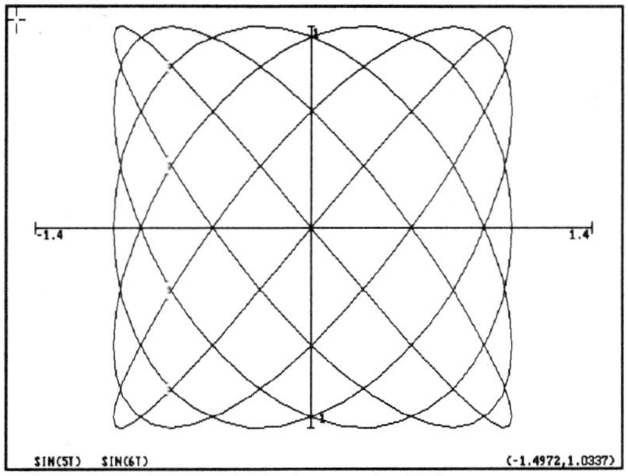

Fig. III.4.6: x=sin(5t), y=sin(6t)

Like polar curves, parametric curves are drawn with 200 line segments, unless you specify another number. The command below draws the cusp $x = t^2$, $y = t^3$, $-1 \le t \le 1$, with only 20 line segments (Fig. III.4.7).

```
Paramg(t^2,t^3; t=-1 to 1
  step 20)
```

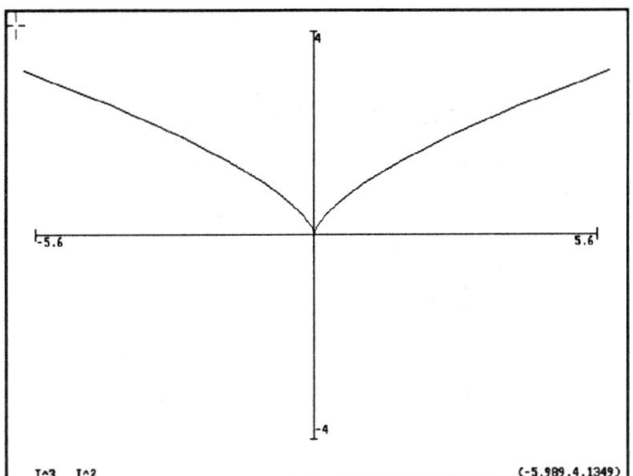

Fig. III.4.7: $x=t^3$, $y=t^2$

III.4.4 Line Segments, Rectangles, Marks and Points

You can place a line segment on the graphics screen from the point (x, y) to the point (z, w) with the command:

```
Line(x,y,z,w)
```

To draw a rectangle with opposite corners (x, y) and (z, w), use the command:

```
Rect(x,y,z,w)
```

You can place a dot at the point (x, y) with the command:

```
Plot(x,y)
```

and you can mark the point (x, y) with a cross by executing:

```
Mark(x,y)
```

A mark at (x, y) can be removed with the command:

```
Unmark(x,y)
```

The last three commands can be applied to lists x and y (III.10). They plot, mark and unmark the points $(x[i], y[i])$.

To leave a mark at the location of the crosshairs, use the command:

```
Mark(CrossX, CrossY)
```

When you execute this command, the crosshairs will jump one pixel up and to the right to avoid obscuring the mark you have just placed on the screen.

The operators `Line` and `Plot` are useful mostly in CC's subroutines. One plotted point is almost invisible, but a group of them can be used to fill a region or trace a curve.

III.4.5 Line Styles and Colors, Labeling Graphs and Filling Regions

Normally, when you draw several graphs successively without erasing, each is drawn in with a different line style (up to four) or, if you are using an EGA or VGA color system, a different color. You can control the line style used to draw the next graph or line by executing the command:

```
Linestyle(n)
```

where n is an integer between 0 and 3, and you can set the color of the next graph or line with the command:

```
Color(ColorWord)
```

`ColorWord` can be one of the built-in color constants `BkColor` (the background color of the graphics screen), `CurrentColor` (the color currently used), or the colors:

`Black`	`Red`	`DarkGray`	`LightRed`
`Blue`	`Magenta`	`LightBlue`	`LightMagenta`
`Green`	`Brown`	`LightGreen`	`Yellow`
`Cyan`	`LightGray`	`LightCyan`	`White`

Different linestyles show clearly only on long line segments. Graphs may consist of individual line segments too short to exhibit the different styles of line drawing.

To prevent CC from automatically changing the linestyle or color of successive graphs, enter the command

```
KeepLine
```

To restore the automatic changes, enter the command

```
VaryLine
```

To erase a curve or line segment from your screen, redraw it using the background color. The following routine will erase the curve drawn by any graphing command:

```
c = CurrentColor   // Remember current color
Color(BkColor)     // Change to background color
  // Move cursor to graphing command and re-execute
  // it, then return here
Color(c)           // restore original color
```

To draw graphs and line segments with thick lines instead of thin ones, enter the command:

```
ThickLine
```

To return to thin lines, enter the command:

```
ThinLine
```

Fig. III.4.8 illustrates all combinations of four line styles and two thickness settings.

A label can be placed at any location on the graphics screen with the command

```
Write@(x,y; LABEL)
```

where (x, y) are the coordinates of the lower-left corner of the label, and `LABEL` can be a formula to be evaluated or a string surrounded by apostrophes. Here are some examples of graphics screen messages:

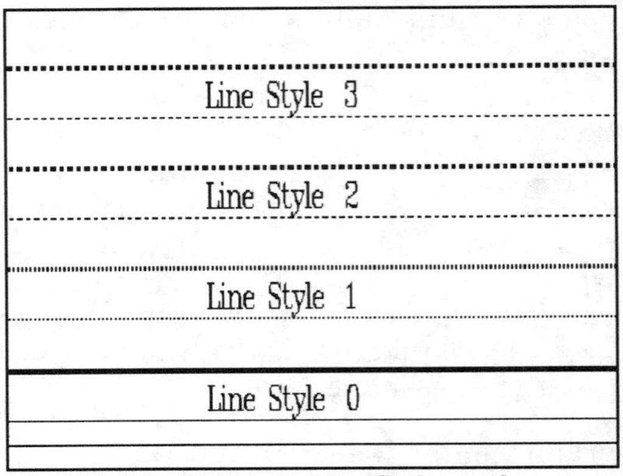

Fig. III.4.8: Different Line Styles and Thicknesses

```
Write@(-1.4,1.4; 'F(x)')
Write@(0.5, 0.5; '(-0.5, 0.5)')
Write@(-1,0.5; sin(π/6))
```

Labels are written in the current color. For more information about labels, see IV.4.1.

The easiest way to place a label accurately on the graph is to move the crosshairs to the point where you want the message to appear (remember, you can return to the last graphics screen by pressing function key F9), press <ENTER> to return to the Input Window, and issue the command:

```
Write@c(LABEL)
```

The lower left corner of the label will be at the crosshairs. For example, to write the coordinates of the crosshairs at their location on the screen, use the command:

```
Write@c((CrossX, CrossY))
```

(Both sets of parentheses are necessary--the label is the list `(CrossX, CrossY)`. See III.10 for information about lists.)

You can increase the size of the labels by adding an optional parameter to the `Write@` and `Write@c` commands. The size parameter can be any integer between 1 and 10, for example:

```
Write@(0.5, 0.5,'F(x)',4)
```

Thicker lines and larger labels are useful for preparing graphs that will be reduced for publication or that will be displayed on an overhead projector.

Labels can be written vertically (bottom to top) as well as horizontally by using the `VWrite@` and `VWrite@c` commands. To write the vertical label "Y-Axis" with size 3 beginning at the current location of the crosshairs, enter the command:

```
VWrite@c('Y-Axis',3)
```

The parameters for `VWrite@` and `VWrite@c` are the same as those for `Write@` and `Write@c` respectively.

You can fill a region, for example a bar on a bar chart or an area inside a curve, with `Fill`. The region to be filled must be surrounded by solid lines (lines drawn with `LineStyle(0)`) of a single color. The form of the `Fill` command is:

```
Fill(x,y,fillcolor,bordercolor,fillstyle)
```

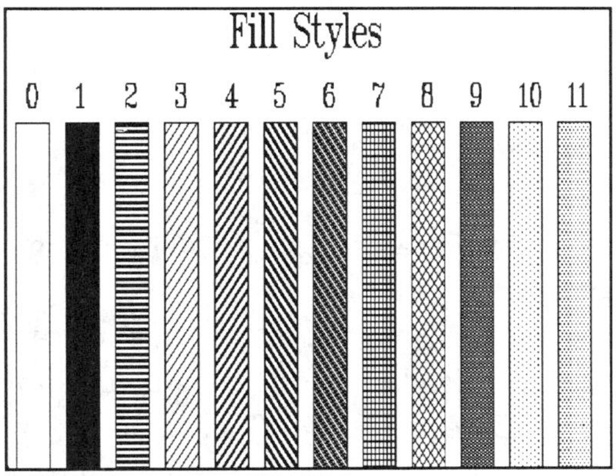

Fig. III.4.9: Fill styles

where `(x, y)` is any point inside the region to be filled; `fillcolor` is the color to fill with, `bordercolor` is the color of the border to fill to, and `fillstyle` is a number between 0 and 11 that indicates which fill style to use. Style 0 means to use "empty fill" or fill with the background color. Fig. III.4.9 shows the different fill styles. It was created with the subroutine `Styles` on the next page (see IV for information on subroutines).

Tip

While you are constructing a complex graphics screen, it is a good idea to save it periodically on disk (III.4.7). That way, if you mess it up, you can recover your partially completed work and continue.

```
Procedure styles
  // Purpose:  Draw bars filled with different fill styles
  AxisOff                            // No axes in picture
  window(0,1,0,1)
  dx = 0.9/36                        // bars are 2dx wide, dx apart
  c = CurrentColor                   // used in Fill command
  graphics
  for i = 0 to 11 do                 // draw 12 bars
      j = (3i+1)dx                   // left edge of bar i
      rect(j,0,j+2dx,.75)            // bar i in color c
      fill(j+dx,0.5,c,c,i)           // filled with style i
      write@(j+2dx/3,0.85,i,3)       // and labeled with i
    end
  write@(.35,1,'Fill Styles',4)      // label above bars
  text
end
```

III.4.6 Three-Dimensional Graphs

CC can sketch and rotate graphs in $\mathbb{R}^3$. The graph can be the graph of a single equation $z = f(x, y)$ or three parametric equations $x = f(u, v)$, $y = g(u, v)$, $z = h(u, v)$. These two types are explained separately, but they have this in common. To construct the graph, you select a rectangular (x, y)- or (u, v)-domain for the variables and a rectangular block in (x, y, z)-space to contain the graph. If you are graphing a single equation, then the (x, y)-domain chosen will also be the base of the (x, y, z) block to be displayed. You must specify the names of the independent variables x and y (or u and v) to use for graphing, since the expressions to be graphed might also include other variable names. You must also select the number n of subdivisions for each side of the domain rectangle.

CC does not automatically graph the entire surface lying over the selected rectangular domain, because it is possible that this surface includes a point at infinity. For this reason, the user must select the block limiting the visual portion of the surface to be graphed.

The graph is constructed by dividing the domain rectangle into n^2 sub-rectangles and mapping the four corners of each sub-rectangle to four points on the surface. The portion of the surface over the sub-rectangle is approximated by a quadrilateral drawn around the four spatial points. The quadrilaterals are then projected onto the screen according to whatever viewpoint is adopted, and the projections are filled and displayed from back to front in order to create a hidden-line effect. A spatial quadrilateral not completely contained within the visible block designated by the user will not be displayed.

For example, to sketch the graph of $f(x, y) = x^2 - y^2$ over the region $-1 \leq x, y \leq 1$ including only values between -1 and 1 and using 100 quadrilaterals to approximate the graph, use the command:

```
Graph3D(x^2-y^2; x=-1,1; y=-1,1; -1,1 step 10)
```

(Fig. III.4.10).

The smallest useful number of subdivisions is usually 10, and the largest number permitted is 40. The time required for graphing is approximately proportional to the square of the step value.

Once a surface appears on the screen, you can rotate it both ways around each axis. To rotate the surface about a vertical axis, press z or Z; to rotate it about a horizontal axis, press y or Y; and to rotate it about an axis pointing out of the screen, press x or X. To speed up multiple rotations, you can press several rotation keys in quick succession. CC will perform all the rotations internally before showing the figure in its final position.

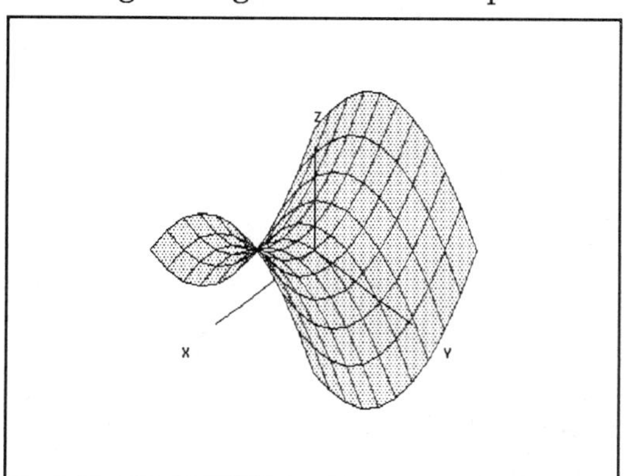

Fig. III.4.10: $z = x^2 - y^2$

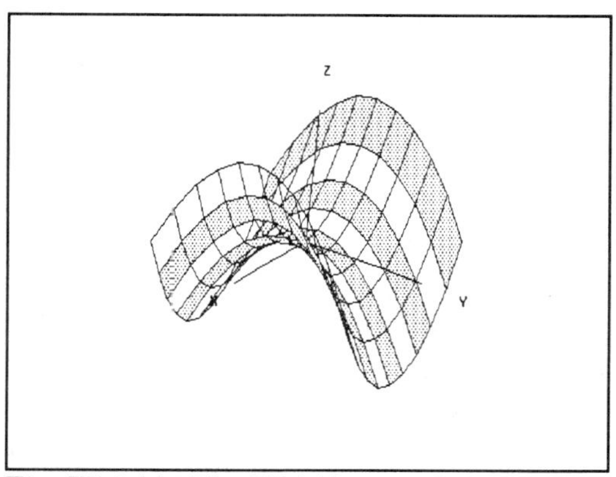

Fig. III.4.11: Fig. III.4.10 rotated

You can also change the shading of a graph from opaque to striped or transparent. To make the figure opaque, press `1` while the figure is on the screen. To make it striped, press `2`; to make it striped the other direction, press `3`; and to make it transparent, press `4`. Fig. III.4.11 shows Fig. III.4.10 rotated by the commands `Y Y Y z z` and shaded with style 3.

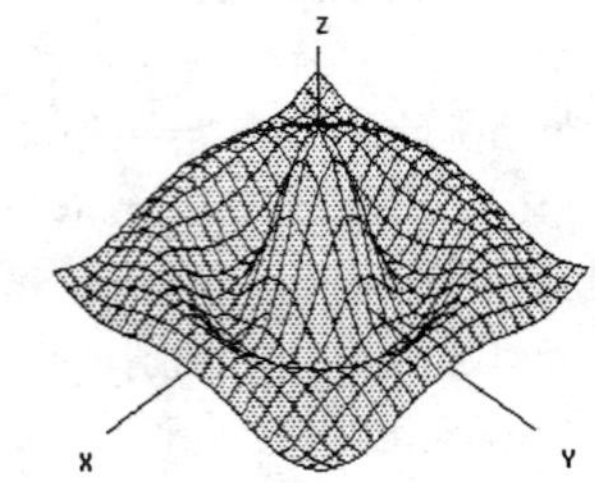

Fig. III.4.12: z = Sin(r)/r

Here is another example, the two-dimensional version of the function sin(x)/x (Fig. III.4.12). It uses user-defined functions, as introduced in II.8:

```
r(x,y)  = sqrt(x^2 + y^2)
g(x,y)  = sin(r(x,y))/r(x,y)
Graph3d(g(x,y); x=-9,9; y=-9,9; -1,1 step 15)
```

III.4.6.1 Parametric Surfaces and Space Curves

CC can sketch surfaces defined by parametric equations. Consider the parametric equations:

$$x = r\cos(\theta)$$
$$y = r\sin(\theta)$$
$$z = \frac{r}{\sin(r)}$$

We will graph the portion of the surface corresponding to the parametric values $0 \le \Theta \le 2\pi$ and $0 \le r \le 9$, and we will display the surface in the box $-9 \le x \le 9$, $-9 \le y \le 9$, $-1 \le z \le 1$. We will also divide the range of each parameter into 15 intervals. All this is accomplished with the command:

```
ParamG3D(r*cos(t),r*sin(t), sin(r)/r;t=0,2*pi; r=0,9; -9,9;
   -9,9; -1,1 step 15)
```

(Fig. III.4.13). The figure has been rotated by pressing Y twice. It is the same surface as appears in Fig. III.4.12 but graphed with cylindrical instead of cartesian parameters.

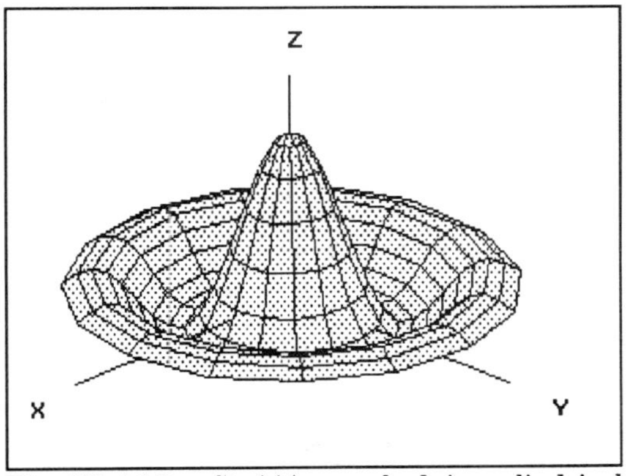

Fig. III.4.13: Sin(r)/r graphed in cylindrical coordinates.

When you type a long example like the command above, do not press <ENTER> until you have reached the end of the formula. It may not break on your screen exactly as it is divided on the page, but if you keep typing until you reach the end, it will work correctly.

Another surface in cylindrical coordinates is $r = 1 + 0.25z^2$; in rectangular coordinates its parametric equations would be:

$$x = (1+0.25z^2)\cos(\theta)$$
$$y = (1+0.25z^2)\sin(\theta)$$
$$z = z$$

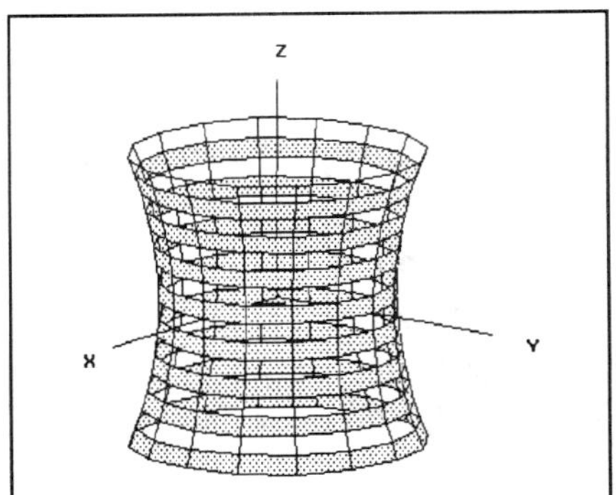

Fig. III.4.14: $r=1+0.25z^2$

The domain for the parameters will be $0 \leq \theta \leq 2\pi$, $-1 \leq z \leq 1$, and the resulting surface will be displayed in a box with limits $-1.5 \leq x \leq 1.5$, $-1.5 \leq y \leq 1.5$, $-1 \leq z \leq 1$. The command to create the graph (Fig. III.4.14) is:

```
ParamG3D((1+0.25*z^2)*cos(θ),
   (1+0.25*z^2)*sin(θ), z;
   θ=0 to 2*pi; z=-1 to 1;
   -1.5,1.5;-1.5,1.5; -1,1 step 16)
```

See VI.1.20 for another approach to the hyperboloid.

Our final parametric surface is the sphere in spherical coordinates (Fig. III.4.15)

```
ParamG3D(sin(φ)*cos(Θ), sin(φ)*sin(Θ), cos(φ); φ=0 to pi,
  Θ=0 to 2*pi; -1,1; -1,1; -1,1 step 16)
```

Space curves can also be drawn with the command `ParamG3D`; just ignore one of the parameters. To draw the helix $x = \cos(t)$, $y = \sin(t)$, $z = t$ (Fig. III.4.16), use the command:

```
ParamG3D(cos(t),sin(t),t; t=-2pi to 2pi;  s=0 to 1;
 -1,1; -1,1; -2pi,2pi step 30)
```

The dummy parameter `s` is necessary.

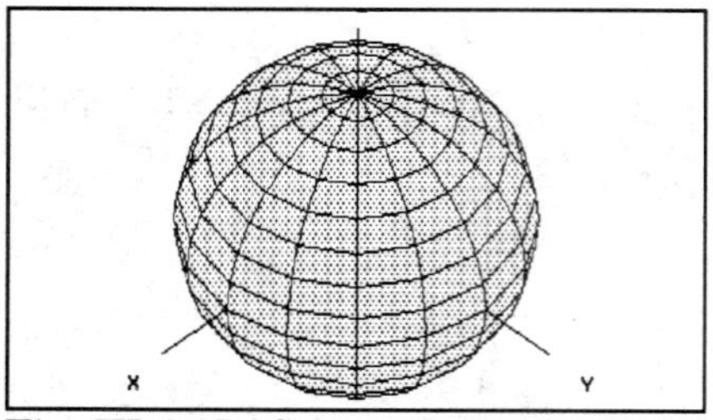

Fig. III.4.15: Sphere in spherical coordinates

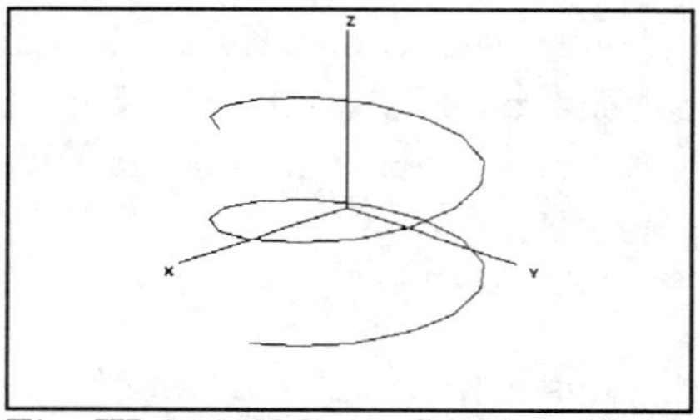

Fig. III.4.16: Helix x=Cos(t), y=Sin(t), z=t

III.4.7 Saving Graphs on Disk

All graphics images are automatically saved to a special graphics buffer from where they can be recalled by pressing function key F9. Depending on how much memory your computer has and the type of graphics system you are using, the graphics buffer can be in memory or on disk. In either case, the graphics buffer stores only the last graphics image. To store additional images for recall to the screen, you have to take specific action.

To save any graphics screen in a disk file for later viewing, press function key `F7` while the screen is displayed. You will be prompted for the name of the file for saving the graphics image. Your name may include a path, or you may accept the default name suggested. If you give the name of an existing file, you will be warned that you are about to replace an existing file and asked if you wish to proceed.

Saving a graphic image does not affect the image on the screen.

To recover a saved graph you must be viewing a text screen, either the Input Window or the Subroutine Window. Press F8, select option (G)raphics, and enter the name of the file containing the graph. The graph will be displayed on the screen as though you had just drawn it, and screen becomes the "Last Graph" when you press <ENTER> to return to the Input Window or the Subroutine Window. Thus you can reload a graphic image and add to it.

When you load a graphic image, you write over the existing "Last Graph." To preserve the "Last Graph" before loading a new one, display it by pressing F9, save it in a file, return to the text screen by pressing `<ENTER>` , and load the new graphic.

Files containing 3D-graphics can be transferred between any two computers that will run CC, but files containing 2D-graphics can only be transferred between computers with compatible graphics systems. That means that both computers use one of the systems: CGA or Hercules or EGA or VGA. Files for 3D-graphics are lists of coordinate values that can be interpreted by CC on any computer, but files for 2D-graphics are actual bit-images of the graphics screens and can only be transferred to a similar graphics system

III.4.8 Printing Graphs

CC's graphics can be printed on any Epson®, IBM® or Hewlett-Packard® LaserJet® compatible printer. Most printers are compatible with one of these three standards. To print a graph, push function key `F2` while the graph is displayed on the screen, and select the option P for (P)rint. The first time you print a graph, you will be asked what type of printer you are using. Respond to CC's query by entering `E` or `I` or `H`. If the graph is not printed correctly on a dot matrix printer with one of the choices `I` or `E`, try the other.

CC can also save graphics images in disk files in a format that can be imported into your word processor or copied to a printer as a stand-alone file. To save a graphics image in one of these files, press F2 when the graphics image is on the screen and select option (D)isk. You will be asked for the file name to use and whether you want to create a file for Epson, IBM or HP-compatible printers.

Graphics files saved for printing are entirely different from those saved for reloading (III.4.7). The graphics screens saved for reloading can be displayed by CC, but they cannot be printed as stand-alone files; the graphics files saved for printing cannot be read back into CC, but they can be printed as stand-alone files or included in word-processing documents. The printable files are text files with imbedded control characters that direct a printer to print the graphic image you have saved. They begin with control characters that put your printer into graphics mode and end with control characters that return your printer to text mode. The last line is a blank line terminated by a line feed-carriage return pair.

The graphics in this manual were created with CC but saved and printed with another screen printing program, namely WordPerfect®'s Grab program. CC's graphic screen printer does not allow for cropping or editing of graphics images. Although it is very handy for making a record of your work, CC's screen printer defers to more sophisticated programs like Grab or Pizazz® or HiJack® for turning CC's graphics into publication quality images.

III.5 Differentiation

CC can compute the derivative of any expression with respect to any variable and evaluate the derivative by substituting a value for the variable. It can also return the symbolic derivative of an expression by omitting the evaluation step. For example, to find the derivative:

$$\left.\frac{d}{dr}\sin(5r)\right|_{r=\pi}$$

enter the command:

```
Dif(sin(5*r), r=pi)
```

This expression returns the value -5, and it can be assigned to a variable or otherwise used like any other numerically valued expression.

Any value whatsoever can be substituted for the variable of differentiation; you are not limited to simple numerical values like π. Consider the expression:

```
Dif(sin(5*x), x=t^2)
```

If t has a numerical value, then a numerical value is returned; if t has a symbolic value, then the square of that value is substituted for x in the expression. In particular, if t is an undefined variable name (a variable name to which no value has been assigned), then this expression returns the symbolic derivative `5*Cos(5*t^2)`.

There is a useful shorthand for defining symbolic derivatives. If x is an undefined variable name, then the command:

```
Dif(f(x),x)
```

returns the symbolic derivative of the expression $f(x)$. Thus, to find the symbolic derivative of Sin($5x$), provided x is an undefined variable name, all you need enter is:

```
Dif(sin(5*x),x)
```

CC returns `5*COS(5*X)`.

In general, the formula `Dif(f(x),x)` means exactly the same as `Dif(f(x),x=x)`, but this last expression must be understood correctly. It means to differentiate the expression $f(x)$ with respect to x, ignoring any value that x might have outside the expression, and then to substitute the value that x has as a free variable for each instance of x in the result. Thus, if x has no prior value, `Dif(sin(5*x),x)` returns `5*Cos(5*x)`. For information about how all this relates to defining one function as the derivative of another, see (III.11.1); for more about variables, see (V.1).

Expressions that include the operators `In` (integral), `Implicit`, `Sum` and `Product` (see III.6, III.7 and III.9) can be differentiated. CC knows the derivatives of these operators. For example, if x is undefined, you can compute the derivative of $\frac{d}{dx}\int_0^{x^2} e^{tx}dt$ by entering the expression:

```
Dif(In(Exp(t*x),t=0 to x^2),x)
```

The result is $\int_0^{x^2} te^{tx}dt + 2xe^{x^3}$, or:

```
In(t*Exp(t*x), t = 0 to x^2) + 2*x*Exp(x^3)
```

CC can only differentiate expressions in which the variable of integration is explicitly included. If you define:

```
a = 3 + Exp(x)
```

where x is an undefined variable, then the expression:

```
Dif(a,x)
```

will return 0, not *Exp(x)* . Only explicit occurrences of `x` are differentiated in `f(x)`. Indirect references to the variable `x` are not recognized. In this way CC differs from symbolic calculators like *Reduce, Mathematica* or *Maples*. CC differentiates functions, not expressions.

III.5.1 Higher-order Derivatives

An optional parameter can be added to CC's differentiation operator to return a derivative of order greater than 1. To find the third derivative of Sin($5x$) at $x = 0.3$, for example, enter the expression:

```
Dif(Sin(5*x),x=0.3,3)
```

Since both the value to substitute for `x` and the order of differentiation can be either specified or omitted, the possible ambiguity in the expression `Dif(Sin(5*x),x,3)` must be eliminated. This expression could mean either `Dif(Sin(5*x),x=3,1)` or `Dif(Sin(5*x),x=x,3)`. To eliminate the ambiguity, the equals symbol = must be used to assign a value to the variable of differentiation, and the comma `,` must be used to set off the order of differentiation. That is:

Entry	Meaning	Value
`Dif(Sin(5*x),x=3)`	`Dif(Sin(5*x),x=3,1)`	5Cos(15) = -3.798
`Dif(Sin(5*x),x,3)`	`Dif(Sin(5*x),x=x,3)`	-125Cos($5x$)

III.5.2 Partial Derivatives

CC will compute the partial derivative of an expression (or function--see III.11) of several variables. For example, to compute the partial derivative of Sin(xy) with respect to y at the point $(x, y) = (3, 4)$, enter the command:

```
Dif(Sin(x*4),x=3)
```

which returns 4·Cos(12) = 3.375. The idea is that you must transform the expression to be differentiated into a function of one variable by substituting values for the other variables before differentiating, and then differentiate with respect to the remaining variable.

Symbolic partial derivatives can also be computed. The expression `Dif(Sin(x*y),x)` returns the value `Y*COS(X*Y)`.

Higher partial derivatives can be computed, but mixed partials require multiple derivatives. To compute $\frac{\partial^3}{\partial x^2 \partial y)}(e^{xy})$ use the formula

```
Dif(Dif(Exp(x*y),x,2),y)
```

III.5.3 Graphing Tangents

CC will graph the tangent line to a curve at a point. To graph the tangent line to the curve y = Sin($5x$) at the point $x = 0.3$, use the command:

```
DiffG(Sin(5*x),x=0.3)
```

(Fig. III.5.1). The command above works exactly like the derivative operator `Dif(Sin(5*x),x=0.3)`; it returns the derivative value 0.353. But it also draws the graph of `Sin(5*x)` and, if the given point (0.3, Sin(1.5)) is on the

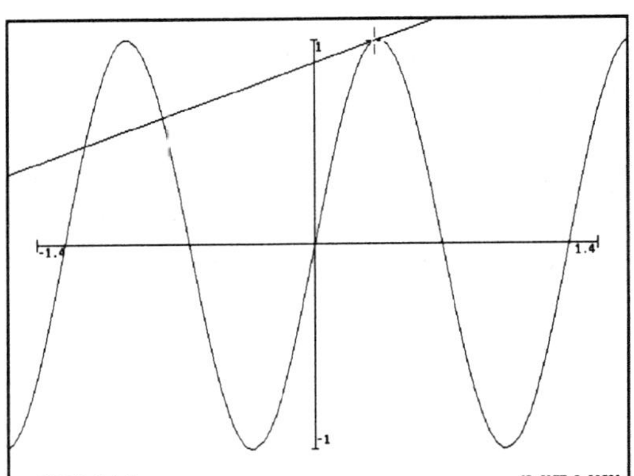

Fig. III.5.1: Tangent line to y = Sin($5x$) at $x = 0.3$

screen, draws the tangent line to the curve at the point. `DiffG` always returns the first derivative of the function. It is intended primarily for classroom demonstrations and other educational purposes.

III.6 Integration

CC will integrate any expression over any finite interval. To compute $\int_0^{\pi} Sin(5x)\,dx$, enter the formula:

```
In(Sin(5*x), x=0 to pi)
```

CC returns 0.4. The integration is carried out numerically using an adaptive Simpson's algorithm (V.3.1). Some examples follow:

Math	*CC*	*Value*
$\int_0^3 \frac{1}{1+x^3}dx$	`In(1/(1+x^3), x=0 to 3)`	1.154449
$\int_0^2 e^{-x^2}dx$	`In(Exp(-x^2), x=0 to 2)`	0.882081

You can speed up the operation of integration, at the expense of accuracy, by replacing `In` with `FIn` (fast integration). Use `FIn` in place of `In` for integrals used in functions that you want to graph. To see how much difference this can make, compare:

```
In(Sin(x), x = 0 to pi)
FIn(Sin(x), x = 0 to pi)
```

To adjust the factors controlling the speed and accuracy of integration more precisely, see III.12.5.

III.6.1 Improper Integrals

CC will attempt to integrate any expression you give it. If an integral calculation does not converge after a specified number of subdivisions, the best answer is returned along with one of two warnings that the answer may not be trustworthy. If CC displays the message:

```
Integral fails to converge--possible improper integral
```

then one of the two approximations that CC uses for the integral was undefined. Usually this means that the integrand was undefined at one endpoint of the domain of integration. If CC displays the message:

```
Possible inaccuracies in integral due to rapid changes in integrand
```

it means that the two approximations used by CC for the integral failed to converge to a single value. Usually this happens when the function being integrated has a vertical tangent.

These warnings do not mean that your answer is necessarily wrong, but when they appear you should consider your integral carefully. You may be able to determine that it is convergent or not by theoretical means, and you may find an alternate way to compute it. One simple method for checking the accuracy of CC's result is to divide the domain of integration into two parts and recompute the integral on each part. To see how this works, consider the following two improper integrals:

Entry	Value
`a1 = In(1/x,x=0 to 1)`	10.20
`a2 = In(1/Sqrt(x),x=0 to 1)`	1.99

Both of these calculations cause CC to display its improper integral warning, although the first is not convergent and the second is. Dividing each integral into two parts and recomputing, we find:

Entry	*Value*
`a3 = In(1/x,x=0,0.1)+In(1/x,x=0.1,1)`	12.5
`a4 = In(1/Sqrt(x),x=0,0.1)+In(1/Sqrt(x),x=0.1,1)`	2.00

If the sum of the parts approximately equals the original integral, then the integral is probably convergent and CC's result is probably approximately correct. Our sample results suggest correctly that the integral of $1/x$ is not convergent and that the integral of $1/\sqrt{x}$ converges to 2.

III.6.2 Graphing Integrals

You can display the region being integrated with the command `InteG`. This command works just like `In`; it returns the value of an integral. It also causes CC to graph the curve being integrated and shade the area of the integral. For example:

```
window(-2*pi,2*pi,-1,1)
integ(Sin(x), x = 0 to pi)
```

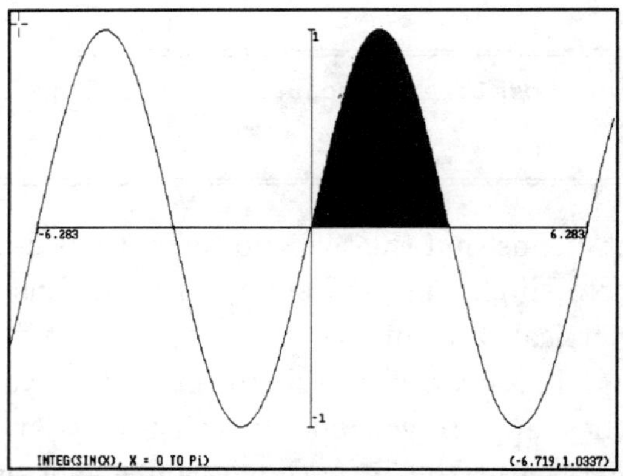

Fig. III.6.1: Integral of sin(x) using INTEG

computes $\int_0^\pi \sin(x)\,dx$ and fills in the area being integrated (Fig. III.6.1). The graphing window must contain the region of integration for this command to work. When CC shades in the region of integration, it actually draws a vertical line for every value it samples while computing the integral. The order in which the lines appear illustrates the adaptive Simpson's method used by CC to compute integrals.

III.7 Solving Equations

To use CC to solve an equation, you must specify the equation and an approximation to the solution. To solve the equation Cos(x)=x, enter the expression:

```
Solve(Cos(x) = x; x = 1)
```

The solution to the equation is stored in the variable `x`, which will be created if it does not already exist. If `x` is defined prior to the execution of `Solve`, then its prior value will be replaced by the solution to the equation. The solution is found by Newton's method starting with the approximation that you enter. Usually the solution found is the one closest to the first approximation.

CC can solve for complex as well as real roots. The expression:

```
Solve(Sin(x) = Cos(x)+2; x = î)
```

returns x = 2.36-0.881i. This result can be checked by entering:

```
Sin(x)-Cos(x)
```

and seeing that its value is nearly 2.

Newton's method for solving equations can fail to find a root if the first approximation is not sufficiently accurate. Consider the equation tan(x) = x ; we wish to find the first positive solution to this equation. First graph the two functions: (Fig. III.7.1):

```
Window(0,3,0,3)
Graph(Tan(x),x)
Graph(x,x)
```

The rightmost intersection in the figure is the desired one. To find the approximate solution to the equation, move the crosshairs to the intersection, return to the text screen, and enter the expression:

```
Solve(Tan(x)=x; x = CrossX)
```

CC will return x = 4.493409. Alterna-

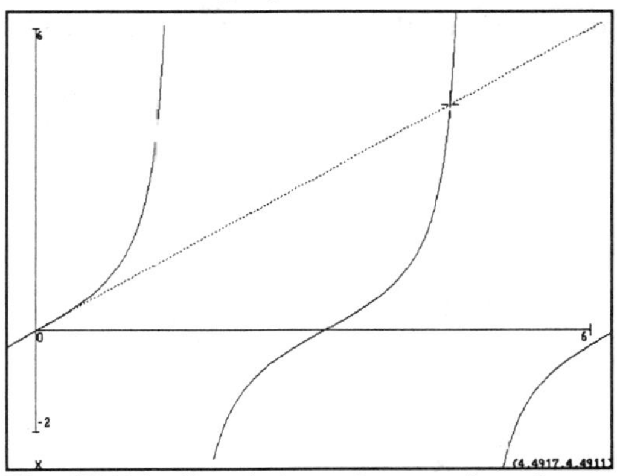

Fig. III.7.1: Crosshairs at intersection of y = tan(x) and y = x

tively, you could use the approximate value 4.5 in place of `CrossX`. If you use a less precise first approximation in this expression, for example `x = 4`, you will find that `Solve` fails to converge to the correct value.

If your equation is extremely complicated, or if you do not need a very precise solution, you can speed up the solving process at the expense of accuracy by replacing `Solve` with `FSolve` (FastSolve). For even finer control of the tradeoff between speed and accuracy in the solving algorithm, see III.12.5 for information about specifying the factor that stops this algorithm.

If the solving algorithm fails to converge satisfactorily, no solution is returned and CC displays the error message:

```
SOLVE or IMPLICIT fails to converge
```

III.7.1 Solving Systems of Equations

The solve operator can be used on systems of equations exactly like it is used on a single equation. Systems with the same number of variables as equations can be solved. You can enter up to eight equations, and you must specify an initial value for each variable. For example, to solve the system:

$$x^2+y^3+z^2 = 3$$
$$x^2-y^2-z^3 = 0$$
$$x^3+y^2-2z^2 = 1$$

for the root nearest (1,1,1), you would enter the line:

```
Solve(x^2+y^3+z^2=3,x^2-y^2-z^3=0,x^3+y^2+2*z^2=1;
  x=1, y=1, z=1)
```

The solve operator will assign the solutions of this system to the variables x, y and z: $x = 1.23$, $y = 0.880$, $z = 0.901$.

CC will solve systems of linear equations quickly and accurately no matter what point you use for a first approximation, but for systems of non-linear equations you should select a first approximation as close as possible to the final solution.

If your system of equations is too long to write as a single entry in CC, you can shorten it by defining each equation as a user-defined function (II.3, III.11) before solving the system:

```
f1(x,y,z) = x^2 + y^3 + z^2
f2(x,y,z) = x^2 - y^2 - z^3
f3(x,y,z) = x^3 + y^2 - 2*z^2
Solve(f1(x,y,z)=3, f2(x,y,z)=0, f3(x,y,z)=1; x=1, y=1, z=1)
```

III.8 Tables of Function Values

It is often useful to produce a table of values for a function, or to create a table comparing two functions. CC will produce such tables with the command `Table`. The command:

```
Table(f(x);x=a to b step c)
```

causes CC to produce a table of values for $f(x)$ using $x = a, a+c, a+2c, \ldots$ up to the last value of x not exceeding b. For example, if you entered:

```
Table(Sin(x); x = 0 to pi/2 step pi/40)
```

CC would produce a table of 21 values for Sin(x) as x ranged from 0 to $\pi/2$. The entry:

```
Table(Sinh(x),Cosh(x); x = 0 to 1 step 0.1)
```

produces a table of 11 values for the functions Sinh(x) and Cosh(x) for $0 \le x \le 1$ (see below).

The command `Table` is particularly useful when used with the printing option (Function key F2). If printing is enabled, then tables displayed on the screen are also printed. In this way, CC can be used to print tables of function values.

X	SINH(X)	COSH(X)
0	0	1
0.1	0.100167	1.005004
0.2	0.201336	1.020067
0.3	0.30452	1.045339
0.4	0.410752	1.081072
0.5	0.521095	1.127626
0.6	0.636654	1.185465
0.7	0.758584	1.255169
0.8	0.888106	1.337435
0.9	1.026517	1.433086
1	1.175201	1.543081

III.9 Sums and Products

CC has operators corresponding to the mathematical operators Σ and $\prod$. Using them, you can define finite sums and products and construct functions like Taylor polynomials or finite Fourier series (III.11.2). The general format for a sum or product is:

```
Sum(f(x), x = a to b step c)
Product(f(x), x = a to b step c)
```

The result is computed by evaluating the expression $f(x)$ at $x = a, a+c, a+2c, \ldots$ up to the last entry in the series not exceeding b, and adding or multiplying the resulting values. The parameter `step c` is optional; if it is omitted, a default step size of 1 is assumed. Here are some sample sums:

Entry	*Value*
`Sum(j; j = 1 to 10)`	$1+2+\ldots+10 = 55$
`Sum(3^j;j=0 to 10 step 2)`	$1+3^2+3^4+\ldots+3^{10} = 66430$

The limits and step size used for sums and products need not be integral. If you wanted to estimate the integral:

$$\int_2^3 \ln(x)\, dx$$

you might use a midpoint approximation with 10 intervals:

```
0.1[ln(2.05)+ln(2.15)+···+ln(2.95)]
```

To perform this approximation with CC, enter the formula:

```
0.1*Sum(ln(t); t = 2.05 to 2.95 step 0.1)
```

the result is 0.909612.

Readers familiar with computer calculations might suspect that CC could miss the last term of this sum due to round-off error encountered in computing the values of the index variable t. As CC generates the values:

$$t = 2.05, 2.15, 2.25, \ldots$$

it is possible that the round-off error inevitably associated with decimal calculations will create a tenth value just a little bit larger than 2.95. To prevent round-off error from eliminating the desired last term of the sum, CC always increases the final index value b by a fraction $c/10000$ of the step value. Referring to the general form of the sum operator above, that means that the summand $f(x)$ is evaluated at the points $x = a, a+c, a+2c, \ldots$ up to the last entry in the series not greater than $b + c/10000$. To see that this complication is necessary, consider the following two formulas:

Formula	*Value*
`Sum(1/1000; j = 1 to 1000) > 1`	`True`
`Sum(1; j=1/1000 to 1 step 1/1000)`	`1000`

The first result shows that adding 1/1000 to itself 1000 times with CC produces a result slightly larger than 1; that is round-off error. The second result demonstrates nevertheless that CC will add up the correct number of terms in a sum with decimal limits 0.001 to 1.000 and step value 0.001. If the second formula took only 999 steps, then the final value would be 999, not 1000.

A sum or product will be empty if the sign of the step value is opposite the sign of the stopping value less the starting value. The empty sum returns 0 and the empty product returns 1. If the limits or step value in a sum or product do not all have numerical values, then the operator returns as its value the original formula after evaluating as many of its components as possible.

III.10 Lists and Arrays

Lists are finite sequences of values. Lists can store statistical data or represent vectors. Both statistical and vector operations on lists are provided. The entries in a list can be anything: constants, expressions with undefined variables, strings, or even other lists.

Lists are similar to the arrays found in most programming languages, but they are extensible without limit like the arrays in APL. Internally they are represented by linked lists, so they do not have to be declared or DIMensioned like arrays in Fortran or Pascal or Basic.

Any variable can be assigned a list value. If a variable a equals a list, then the Output Window will display a component by component. You will see entries for $a[1]$, $a[2]$, ... If one of the components, say $a[3]$, is itself a list, then its entries will be displayed individually as: $a[3,1]$, $a[3,2]$, ... The first index in a list is always 1. If the list is too long to be displayed in the Output Window, then its values can be scrolled through the window with the function keys F5 and F6.

III.10.1 Entering and Modifying Lists

To define a list, put its elements between parentheses:

```
a = (1,2,3)
b = (x,y,z)
c = ((1,2),3,(4,5))
```

You can also define lists with the `List` operator:

```
List(g(t), t = a to b step c)
```

For example:

```
f = List(t^2, t=0 to 1 step 0.1)
```

results in $f[1] = 0$, $f[2] = 0.01$, $f[3] = 0.04$, ... , $f[10] = 0.81$, $f[11] = 1$.

The third way to construct a list is to concatenate two existing lists by writing the concatenation operator ¦ between them. To concatenate lists a and b, enter:

```
g = a ¦ b
```

The concatenation operator permits adding an element to the beginning or end of a list. To add 2 to the beginning of a list a, enter:

```
2 ¦ a
```

to place 2 at the end of the list a, enter:

```
a ¦ 2
```

If a list a is already defined, you can change one of its elements simply by assigning it a new value:

```
a[3] = 7
```

However, you cannot use this method to add an element to a list. To do that, you must use concatenation. If a has only 3 entries, then the assignment:

```
a[4] = 1
```

will not make 1 the fourth element of the list a. Instead, this operation will result in an error, because a has no fourth entry to change to 1. To add 1 as the fourth element of this list, use the formula:

```
a = a ¦ 1
```

An empty list can be created with the assignment:

```
a = ()
```

but a list of length 1 cannot be entered so simply. The expression a = (3) is the same as a = 3; it assigns a the value 3, not the list of length one whose only entry is 3. If you need a list of length 1, adjoin the desired element to the empty list. The expression:

```
a = () ¦ 3
```

results in a list a of length 1 with $a[1] = 3$.

Had CC used braces or some other special symbol in place of parentheses for list delimiters, then lists of length one could have been distinguished from simple values. However this would lead to other difficulties. It would not have been possible to use square brackets for list delimiters, because it is necessary to distinguish lists from sublist selectors. If square brackets were used for both lists and sublist selectors, then the expression `[3,4,5][2]` would be ambiguous. It might mean either the second element of `[3,4,5]` or the product of the two lists. Using braces to denote lists would be unambiguous but unconventional. It seemed more trouble to write `{3,4,5}` for `(3,4,5)` than to muddy the distinction between singleton lists and elements.

III.10.2 Accessing Lists

To produce the j'th element in a list a, use the formula $a[j]$. The first element of a is always $a[1]$. If $a[j]$ is itself a list, then its k'th element can be written either as $a[j][k]$ or $a[j,k]$. Here are some sample entries demonstrating list operations:

Formula	*Value*
`a = ((1,2),3,(4,5))`	list defined
`a[1]`	`(1,2)`
`a[1,2]`	`2`
`a[4]`	not defined
`a[2] = (6,7,8)`	`a[2]` redefined
`a`	`((1,2),(6,7,8),(4,5))`

An entire sublist of a list a can be accessed with the expression $a[j{:}k]$, which returns the list:

$$(a[j],\ a[j+1],\ \ldots\ ,\ a[k])$$

For example, if a = (3,4,5), then a[2:3] = (4,5). Note that the first element of the list $a[j{:}k]$ is the element $a[j]$, which is different from the list $a[j{:}j]$. CC distinguishes between elements and list of length 1 (even when the element is itself a list--((1,2)) is a list of length 1 whose only element is the list (1,2)). The expression $[i]$ is called an *element selector*; $[i{:}k]$ a *sublist selector*.

Sublist selectors can be combined like element selectors. Analogous to the element $a[j,k]$ is the sub-sub list $a[j{:}k,l{:}m]$. The formula a[j:k,l:m] returns the list:

$$(a[j,l{:}m],\ a[j+1,l{:}m],\ \ldots\ ,\ a[k,l{:}m])$$

where each of the expressions $a[t,l{:}m]$ is a list:

$$(a[t,l],\ a[t,l+1],\ \ldots\ ,a[t,m])$$

Think of a as a matrix and $a[j{:}k,l{:}m]$ as a submatrix.

Now we come to a narrow distinction. Although $a[i,j] = a[i][j]$, $a[j{:}k,l{:}m] \neq a[j{:}k][l{:}m]$. This last expression is equal to $a[j+l\text{-}1{:}j+m\text{-}1]$.

Sublist selectors can be mixed with element selectors. Here are some sample calculations using sublist and element selectors:

Formula	*Value*
`a = ((1,2,3),(3,4),(6,7))`	list defined
`a[2,1]`	`3`
`a[1:2,1:2]`	`((1,2),(3,4))`
`a[1]`	`(1,2,3)`
`a[1,2:3]`	`(2,3)`
`a[2:3]`	`((3,4),(6,7))`
`a[2:3,2]`	`(4,7)`
`a[2:3][2]`	`(6,7)`

Although you can replace one element of a list by using an assignment statement like `a[2] = 3`, you cannot replace an entire subrange of a list the same way. If a is a list of length 8 and you wish to remove the second and third entries and replace them with 7,8,9, use the formula:

```
a = a[1] ¦ (7,8,9) ¦ a[4:8]
```

III.10.3 List Operations

Lists can be added, subtracted, multiplied and divided by other lists or by constants, and ordinary scalar functions like `Sin` and `Sqrt` can be applied to lists. When two lists are combined, they are combined term-by-term, and the operation stops when the shortest list is exhausted. A scalar combined with a list, on the other hand, is combined with each element of the list. Similarly, a scalar function applied to a list is applied to each element of the list. Here are some examples:

Formula	*Value*
`a = (1,2,3)`	list defined
`b = (4,5,6)`	list defined
`a+b`	`(5,7,9)`
`a/b`	`(0.25, 0.4, 0.5)`
`3*a`	`(3, 6, 9)`
`b-4`	`(0, 1, 2)`
`Sqrt(a)`	`(1, 1.414, 1.718)`
`c = (7,8)`	list defined
`a+c`	`(8, 10)`
`a+c[1:1]`	`(8)` add a to a list of length 1
`a+c[1]`	`(8, 9, 10)` add a to a scalar

The last two lines show how a list of length one differs from a simple value.

In addition to the scalar functions that operate on the entries in a list, there are special list functions that operate on entire lists. For the examples, a = (1,2,3).

Function	*Meaning*	*Example*	*Value*
`Sum`	Sum elements	`Sum(a)`	6
`Product`	Multiply elements	`Product(a)`	6
`Length`	Number of elements	`Length(a)`	3
`Max`	Maximum element	`Max(a)`	3
`Min`	Minimum element	`Min(a)`	1

III.10.3.1 Vector Operations

Lists can be used to represent vectors. The list a = (3,4,5) can be thought of as the vector (3,4,5) in $\mathbb{R}^3$. As explained above, such vectors can be added or multiplied by scalars (numbers). You can construct linear combinations of these vectors, and you can use the list operations to construct the dot product of two vectors or the length of a vector. Here are some sample calculations:

Formula	*Value*
`a = (1,2,3)`	list defined
`b = (4,5,6)`	list defined
`3*a - 4*b`	`(-13, -14, -15)`
`Sum(a*b)`	32 dot product of `a` and `b`
`Sqrt(Sum(a*a))`	3.74 norm of `a`

III.10.3.2 Lists and Statistics

If a is a list, you can find the average and standard deviation of the entries in a with the commands:

```
Average(a)
StanDev(a)
```

If a and b are lists, you can compute the linear regression through the points ($a[i]$, $b[i]$) with the command:

```
r = Regr(a,b)
```

The output of `Regr` is a list r with three elements: $r[1]$ is the slope of the regression line, $r[2]$ is its y-intercept, and $r[3]$ is the correlation coefficient.

Here's a sequence of commands that illustrates CC's statistical operations.

Formula	*Value*
`f() = Random+random+random+random+random`	function defined

$f()$ is a function with no arguments that returns random values with a roughly normal distribution of mean 2.5 and variance 5/12. See III.11 for more information about defining functions.

`a = List(f(),i=1,100)`	100 random values
`Average(a)`	their average
`StanDev(a)`	their standard deviation

Next we will construct some points randomly varying from the line $y = x$ above the interval [0,1], and we will find the regression line through these points.

`a=List(x,x=0 to 1 step 0.1)`	(0, 0.1, 0.2, ..., 1.0)
`b=List(a[i]+f()-2.5,i=1 to 11)`	11 random values normally distributed around the line y=x at x = 0, 0.1, ..., 1
`r = Regr(a,b)`	regression through points (0, b[1]), (0.1, b[2]), ..., (1, b[11])
`r[1]`	slope of regression line
`r[2]`	y-intercept of regr. line
`r[3]`	coefficient of regression

III.10.4 Graphing Lists

If a and b are lists, you can connect the sequence of points ($a[j]$, $b[j]$) with line segments by using the command:

```
ListG(a,b)
```

You can also place a cross at each of the points ($a[j]$, $b[j]$) by executing:

```
Mark(a,b)
```

and you can erase the crosses with the command

```
Unmark(a,b).
```

You can use the list graphing commands to illustrate the regression calculation above:

`Window(0,1,-1,2)`	Window to display points
`KeepLine`	Use only solid lines for graphing
`Mark(a,b)`	Mark the points $(a[i], b[i])$
`ListG(a,b)`	Connect these points
`ThickLine`	Use thick line for next graph
`Sketch(r[1]*x+r[2],x)`	Plot regression line

Fig. III.10.1 shows the result of these operations.

To display a histogram of the elements in a list *c*, use:

```
Histogram(c,r,s,n)
```

CC divides the interval $[r, s]$ into n subdivisions and shows the histogram of the elements of c according to these subdivisions. Outliers below r or above s are included in the bottom and top categories respectively. If c is a list of integers in the range $1..n$, you can graph its histogram with the command:

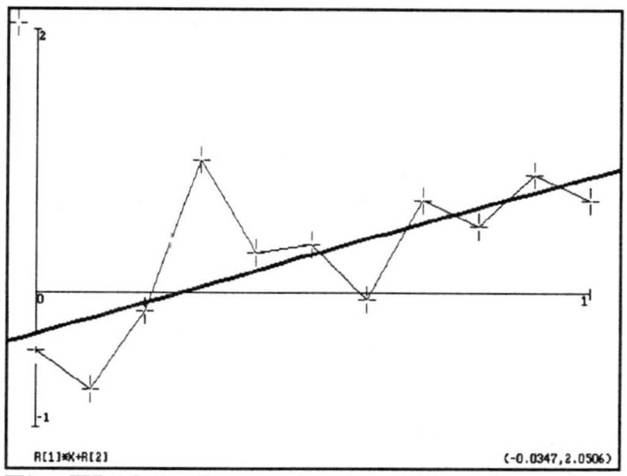

Fig. III.10.1: Regression line through random points

```
Histogram(a,1,n,n)
```

The command `Histogram` is a function that, in addition to graphing the histogram, also returns a list of the frequencies graphed. Here is a sample CC calculation demonstrating the use of histograms. Its graphical output is shown in Fig. III.10.2.

```
f() = random+random+random+random+random
c=List(f(),i=1,100)
Hide(c)
freq = Histogram(c,0,5,10)
```

At the end of this calculation, *freq* = (0, 0, 6, 14, 27, 29, 17, 7, 0, 0). Of course, if you run it, you will get a different result and a different histogram because the function `random` gives different results each time.

If you just want the numerical output from `Histogram` without the graph, you can use the command:

```
Frequencies(c,r,s,n)
```

which uses the same parameters as `Histogram` and returns the same result but does not create the graph.

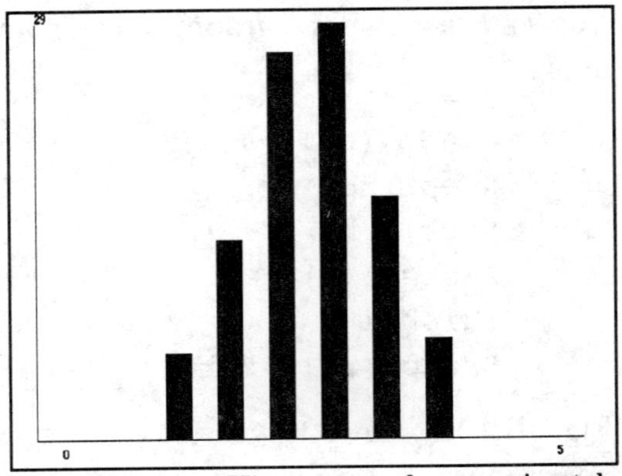

Fig. III.10.2: Histogram of approximately normal random data

III.10.5 Displaying and Printing Lists

Once they are created, lists can be viewed or printed just like any other variables. You can review the values in a list by scrolling through them in the Output Window with function keys F5 and F6. You can print the values in a list ccat least three ways. One way is to turn on the printing option with function key F2, option 1 (*Start Printing*), and execute the variable name (see III.12.1). Another way is to print all of your variables by pressing F2 and selecting option 6 (*Print Variables*). A third method is to select F2, option 8 (*Save One Variable to Disk*) and select `PRN` as the file to save to.

There is a difference between the last two methods. If you select option 6, your variables will be printed according to the current decimal precision setting (III.12.4) with their names, and any symbolic values will be printed as they would appear in CC. If you select option 8, the variable values will be printed in scientific notation without names, and symbolic values will be changed to `UNDEFINED`. Option 8 is really designed for saving variable values to disk, and using it to print the variable is trick you play on CC.

If you want to suppress the display of a long list `v`, enter the command:

```
Hide(v)
```

where `v` is the name of the list you do not want displayed. To restore the variable to the display, issue the command:

```
Show(v)
```

Hidden lists are still defined. They can be accessed in all the usual ways. To delete a variable, you must use the command `Forget`. You can hide or show more than one list simply by writing multiple lists in the `Hide` or `Show` command: `Hide(u,v,w)`.

III.10.6 Saving and Loading Data from Disk Files

A list can be saved on disk in an ASCII file that can be read back into CC or accessed by another program. To save a list, press F2 and select option 8 (Save One Variable to Disk). Tell CC which variable you want to save and the name of the file to save to. The numerical values stored in your variable will be saved in an ASCII file, one value per line, in scientific format. Any symbolic values will be changed to `UNDEFINED`.

To load an ASCII file of numerical data from disk into a list variable, press F8, then "V" to select the option *one (V)ariable*. Tell CC the name of the file you will read and the variable to store the input data. The input file must be an ASCII file with one entry per line. The entries can be any numerically-valued CC expression: numbers in decimal or scientific format (`3.26, 2.4E7`), numerically-valued expressions (`Cos(3.4)`), or even complex values (`3.2-4.3î`). Lists are not permitted. The files created as in the paragraph above can be reloaded back into CC this way.

Files of CC commands can also be saved and reloaded. See III.12.3.

III.11 User-defined Functions

CC enables you to define your own functions. Just type them in the Input Window as you would ordinarily write them:

```
f(x) = sin(2*x)
g(x,y) = x^2 - y^2
```

These functions are called *user-defined functions*, to distinguish them from the built-in functions described in III.2 and from the subroutine functions introduced in Chapter IV. User-defined functions can be evaluated, composed, differentiated, integrated and solved, as in these examples:

Formula	*Value*
`h(x) = g(f(x-1),f(x+1))`	Function defined
`a = h(3)`	a = 0.254
`b = dif(g(3,y),y=4)`	$b = -8 = \left.\frac{\partial(x^2-y^2)}{\partial y}\right\|_{(x,y)=(3,4)}$
`c = in(f(x),x = 0, pi/4)`	$c = 0.5 = \int_0^{\pi/4}\sin(2x)\,dx$
`solve(f(x)=x, x=1)`	x = 0.947747

Function names, like variable names, must consist of no more than twelve alphanumeric characters and start with a letter. If you define a new function with the same name as one already defined, the new definition will replace the old one. On the other hand, if you try to define a function using one of your variable names, CC will give you an error message and not accept the definition. It is also an error to define a variable using a name already used for a function. If you need to replace a variable with a function of the same name or *vice versa*, first use `Forget` to delete the item you wish to replace.

Names used in function definitions are stored as written; they are not evaluated until the function is, and they are re-evaluated each time the function is invoked. If a function includes undefined names when it is evaluated, the value of the function will be a formula with undefined terms. This is best illustrated with some examples:

Formula	*Value*
`Forget(a,t)`	a and `t` become undefined variables
`f(x) = Sin(a*x)`	Function defined
`f(t)`	`SIN(A*T)`
`f(2)`	`SIN(2*A)`
`a = 3`	a is defined
`f(t)`	`SIN(3*T)`
`f(2)`	-0.2974 = Sin(6)
`a = 4`	a is redefined
`f(t)`	`SIN(4*T)`
`f(2)`	0.9893 = Sin(8)

A function can be erased with the command `Forget`, just like a variable. The commands `Hide` and `Show` have no effect on functions.

You can define a function with no variables:

```
f() = random+random+random+random+random
```

Parentheses must be used to make *f*() a function that is re-evaluated every time it is invoked. Each time you execute *f*(), a new random value is generated. Omitting the parentheses would have created a random value and stored it in a variable called *f*. Then, every time *f* was used, it would have returned the same value. Once defined, *f*() can be invoked either as *f*() or *f*. (The example is an approximation to normally distributed random variable with mean 2.5 and variance 5/12.) Chapter V has more information about the difference between functions and variables and how each is evaluated.

III.11.1 Functions with Derivatives and Integrals

The derivative and integral operators `Dif` and `In` can be included in function definitions. For example, if `f(x)` is already defined, you can define `df(x)` to be the derivative of `f(x)` by entering:

```
df(x) = Dif(f(x),x)
```

You could define your own natural logarithm function with the line:

```
mylog(t) = In(1/s, s = 1 to t)
```

You cannot use `Graph`, `QuickG`, `Sketch`, `PolarG`, `ParamG`, `Graph3D` or `ParamG3D` in a function. You can use the operators `DiffG` and `InteG` in a function; they return values just like `Dif` and `In`, but they also cause a graph to be drawn when the function is executed.

III.11.2 Functions with Sums and Products; Taylor Polynomials and Fourier Series

The operators `Sum` and `Product` can be used to define functions. For example, to define the Taylor polynomial of degree $2n+1$ for the function $\sin(x)$:

$$p(x,n) = \sum_{j=0}^{n} \frac{(-1)^j x^{2j+1}}{(2j+1)!}$$

you could use:

```
p(x,n) = sum(par(j)*x^(2*j+1)/(2*j+1)!; j = 0 to n)
```

The Taylor polynomial of degree n for the hyperbolic cosine is:

$$csh(x,n) = \sum_{j=1}^{\lfloor n/2 \rfloor} \frac{x^{2j}}{(2j!)}$$

In CC:

```
csh(x,n) = Sum(x^j/j!; j = 0 to n step 2)
```

CC's `step` parameter makes `Sum` more flexible than the mathematical Σ.

Here are some more examples.

(1) The general Taylor series about a for a function $f(x)$ is:

$$T(x,a,n) = \sum_{i=0}^{n} \frac{f^{(i)}(a)}{i!}(x-a)^i$$

This general form is entered into CC with the formula:

```
T(x,a,n) = Sum(Dif(f(x),x=a,i)*(x-a)^i/i!,i=0 to n)
```

Once you define a function $f(x)$, the function $T(x,a,n)$ will equal the Taylor polynomial of $f(x)$ about $x = a$ of degree n.

(2) The truncated Fourier series for a square wave:

$$sqrw(x,n) = \sum_{j=0}^{\lfloor (n-1)/2 \rfloor} \frac{\sin((2j-1)x)}{2j+1}$$

becomes:

```
sqrw(x,n) = Sum(Sin(j*x)/j; j = 1 to n step 2)
```

(3) You could use `Product` to define the binomial coefficient function $\binom{n}{m}$:

```
b(n,m) = Product((n-j)/(m-j); j = 0 to Min((m-1,n-m-1))
```

(But you do not have to; `Binom(n,m)` is a built-in function in CC.)

III.11.3 Functions of Lists

When CC's built-in scalar functions like `Sqrt` are applied to a list, they "go inside" the list and operate on each element of the list. On the other hand, CC's list functions such as `Sum` operate on entire lists. When a user-defined function receives a list as an argument, it operates like a list function in that it does not automatically operate on the elements of the list but instead attempts to operate on the list as a whole. Of course, it is up to the user to supply a user-defined function with arguments that are appropriate for the definition of the function. Here are some examples of user-defined functions that operate on lists:

Function	*Meaning*
`dot(x,y) = Sum(x*y)`	dot product of two vectors x and y
`norm(x) = Sqrt(dot(x,x))`	Euclidean norm of a vector x
`var(x) = StanDev(x)^2*Length(x)/ (Length(x)-1)`	population variance of x

User-defined functions can also return lists as values. Here are some examples:

Function	*Value*
`r(n) = list(random; j = 1 to n)`	returns n random numbers
`s(n) = list(t, t = 0 to 1 step 1/n)`	$(0, 1/n, 2/n, ..., 1)$
`cross(x,y) = (x[2]*y[3]-x[3]*y[2], x[3]*y[1]-x[1]*y[3], x[1]*y[2]-x[2]*y[1])`	cross product of x and y

III.11.4 Defining Functions Implicitly

Generally speaking, an equation in two unknowns can be said to define either of the unknowns as a function of the other. For example, the equation

$$x^2 + rx - 3 = 0$$

defines x as a function of r. You might have noticed that, strictly speaking, this statement is incorrect. The equation has *two* solutions for each value of r. However, if we begin with an approximate value (initial guess) for x, we can choose x to be the single solution closest to the selected approximate value. In this way, the equation plus the initial approximation define x as a function of r.

To define the function $f(r)$ implicitly in CC, use the operator `Implicit` like this:

```
f(r) = Implicit(x^2+r*x-3 = 0, x = 2)
```

The operator `Implicit` uses the same arguments as the operator `Solve`, but `Solve` is a procedure that places its results in specified variables while `Implicit` is a function (like `Sin`) that returns its result as a value.

If you now enter:

```
a = f(2)
```

you will get the result $a = 1$. You can treat `f(r)` just like any other function of one variable. You can graph it, integrate it, differentiate it, tabulate it, even solve it. Note however that when CC graphs or integrates an expression including `Implicit`, it must evaluate the expression many times over. Since `Implicit` computes much more slowly than a function like `Sin`, you should use `Sketch` to draw graphs and `Fin` to calculate integrals that involve `Implicit`. You can also speed up the operation of `Implicit` itself at the expense of accuracy by changing the factor `SolveTolerance` (see III.12.5).

`Implicit` can be used to define functions of more than one variable, for example:

```
g(r,s) = Implicit(x^2+r*x + s = 0, x = 2)
```

`Implicit` can also define implicit functions from systems of equations. If `Implicit` is applied to a system of n equations in n unknowns, it returns a *list* of the n solutions. Consider, for example:

```
g(r,s) = Implicit(x*y = r, x - y = s; x = 1, y = 1)
```

If you enter:

```
a = g(3,2)
```

CC returns $a[1] = 3$, $a[2] = 1$. The function g(r,s) is a vector-valued function of two variables.

As a special case, the operator `Implicit` can be used to solve systems of equations without bound variable parameters. Here are two examples:

Formula	*Value*
`a = Implicit(x^2+2*x-3, x = 2)`	1
`b = Implicit(x*y = 6, x-y =1; x=1, y=1)`	(3,1)

III.11.5 Differentiating Implicit Functions

Functions defined implicitly can be graphed, integrated, tabulated, solved and differentiated just like any other functions, although implicit functions may be slow to evaluate. This section will explain the form of the derivative of an implicit function. Two examples will be explained, leaving it to the reader to extrapolate to the general case.

The derivative of an implicit function is another implicit function. The first example we will consider is the one above:

```
f(r) = Implicit(x^2 + r*x - 3 = 0, x=2)
```

Computing $f'(2)$ requires no special commands. Just enter:

```
a = Dif(f(r),r=2)
```

CC returns a = -0.25. The ambitious reader may solve the implicit equation for $x = f(r)$ explicitly and differentiate it to check this result.

If r is an undefined variable and you enter:

```
Dif(f(r),r)
```

CC will return the formal derivative of $f(r)$ with respect to r:

```
IMPLICIT(-3+X^2+R*X = 0, X + 2*X*ZZZ1 + R*ZZZ1 = 0,
 X = 3, ZZZ1 = 0)[2]
```

It is the form of this derivative that we wish to explain. The derivative is the second solution to a system of two equations in two unknowns X and $ZZZ1$ and one parameter R:

$$X^2 + R \cdot X - 3 = 0$$
$$2 \cdot X \cdot ZZZ1 + R \cdot ZZZ1 + X = 0$$

This system of equations is just the original equation defining *X* as a function of *R* together with the derivative of this equation with respect to *R*. The variable *ZZZ1* is the derivative of *X* with respect to *R*. The element selector [2] means that the derivative will return the second part of the solution to the system, which is the value of *ZZZ1* or the desired derivative.

For our second example, we use the function of two variables above:

```
g(r,s) = Implicit(x*y = r, x - y = s; x=1, y=1)
```

First we will calculate the derivative of *g(r,s)* with respect to *r* at the point (r,s) = (3,2):

```
a = Dif(g(r,2),r=3)
```

The result is $a[1] = 0.25$, $a[2] = 0.25$.

There are two results because *g(r,s)* is a function that returns a list of two values, so it is really two scalar functions. Its derivative with respect to *r* is two scalar derivatives of two scalar functions. To see the formal derivative of *g(r,s)*, make sure that *r* and *s* are undefined and enter:

```
Dif(g(r,s),r)
```

The result is:

```
IMPLICIT(X*Y = R, X-Y = S, -1+Y*ZZZ1+X*ZZZ2 = 0,
 ZZZ1 + (-ZZZ2) = 0; X = 1, Y = 1, ZZZ1 = 0, ZZZ2 = 0)[3:4]
```

That is, the formal derivative of *g(r,s)* is a system of four equations in four unknowns, from which we will select the last two solutions *ZZZ1* and *ZZZ2*:

$$
\begin{aligned}
&X \cdot Y = R \\
&X - Y = S \\
&ZZZ1 \cdot Y + X \cdot ZZZ2 = 1 \\
&ZZZ1 - ZZZ2 = 0
\end{aligned}
$$

These equations are just the two original equations together with their derivatives with respect to *R*. *ZZZ1* and *ZZZ2* are the derivatives of *X* and *Y* with respect to *R*.

The largest number of equations that can appear in a system of equations in CC is 8, so the largest number that can appear in a function to be differentiated is 4.

III.12 Additional Features of CC

In this section, we discuss printing output from CC, establishing configuration files, CC files on disk, controlling the accuracy of CC's operators and output, and using shorthand abbreviations and Greek letters in CC's formulas.

III.12.1 Printing the Output

CC will create a printed record of all your entries to CC together with the results of CC's calculations and copies of all your graphs. You can leave your printer on while you work and create a record of all your input and output. To start printing, push function key F2 and select menu option 1 (*Start Printing a record of all input and results*). You will be asked to specify the printer you are using: Epson® compatible, IBM-Proprinter® compatible, Hewlett-Packard LaserJet® compatible or none of these. If you are not sure about your printer and you are using a dot-matrix printer, try Epson or IBM; if you are using a laser printer, try HP. If your printer matches one of these choices, then all your text will print correctly, even text with Greek characters and special symbols like √, and you will be able to print your graphs. If your printer does not match one of these, then you will have to select the option "none." If you choose this option, then only ASCII text will be printed. Special characters like π and √ will be replaced by text equivalents, and graphics will not be printed. The only advantage to selecting "none" is that almost all printers will work with this option.

Once you have selected a printer, CC will use it for all of its printing operations until you change it. To select a different printer, press function key F2 and select option 9 (*Change Printer Type*).

To stop printing the record of your entries and results, press function key F2 again and select option 2 (*Stop Printing*).

In addition to making an on-going record of your work with CC, you can at any time print all of the current entries in the Input Window, the values of all the variables or all of the subroutines. Push function key F2, then select from the menu one of the options:

3. *Print all formulas from input window, all subroutines and all variables*
4. *Print all formulas from input window*
5. *Print all subroutines*
6. *Print all variables*

If you have not yet specified which type of printer you are using, you will be queried for one after selecting one of these options.

While you are printing, the screen will display a message indicating that the printer is working. If nothing is happening on your printer, check to make sure it is switched on, supplied with paper and on-line. Eventually the printer will finish printing or--if it is not working--a message will appear on the screen warning you check your printer and asking if you want to try to print again or skip printing and continue calculating. If you skip printing while recording a printed record of your work, the recording feature will be turned off. You can turn it on again by pressing F1 and selecting Option 1.

As explained in III.4.8, you can print any graph by pressing F2 while it is on the screen. However, if you are printing your work, you need not take special action to print each graph. When you leave the graph by pressing <ESC> or <ENTER>, you will be asked if you wish to print the graph.

You can also save an ASCII file of all of your variables with their current values, all of your subroutines and all the entries in the Input Window. These files are identical to the printed version of this information described above. To save the Input Window, subroutines and variables, push function key F2, select Option 7 (*Save input window, subroutines and variables to disk*) and enter a file name for saving. This file cannot be read back into CC, but it is useful for transferring CC's calculations to written reports.

III.12.2 Text Screen Colors

If you are using a color monitor, you can select the text and background color to be displayed. Using the color words found in III.3.4.6, change the screen colors with the command:

```
TextColor(foreground, highlight, background)
```

for example `TextColor(yellow, red, black)`. The background colors must be selected from the first eight colors in the color list.

If you want to save your text colors so that CC will always start with these colors, create a configuration file as explained in III.12.8.

III.12.3 Saving and Loading Files from Disk

Anytime you wish, you can save your calculator entries, including your subroutines (see Chapter IV), in a disk file that can be read by CC. You can use this feature for:

- saving work to be resumed;
- developing subroutine libraries;
- producing a record of calculations for showing in seminars and public addresses;
- writing interactive lessons using CC.

The demonstration file DEMO.CC (II.2) was created using the file creation feature of CC.

To save all of the formulas in the Input Window and the Subroutine Window to a disk file, push function key F7. You will be asked for the name of a file for saving your entries, and a default file name will be suggested. You can either enter a name of your own choosing or accept the default name by pressing `<ENTER>`. Your file name can include a disk or path identifier. If you use the name of a file that already exists, CC will ask you if you want to replace it with your new file. If you answer "Y" for yes, CC will create a record of all your entries in the specified file.

To read saved entries back into CC, push function key F8. You will be offered a menu of three choices. Select the option *(C)alculator entries*. You will be asked for the name of the file you want to read, and the name of the last file used will be suggested as a default name. Enter the name of the file you want to read, or just press <ENTER> to accept the default name. Your file name can include a disk or path identifier.

You can load a file of saved entries when you start CC by adding the name of the file to the command `cc` that you enter to start the program. For example, to start CC with the demonstration file, enter the command

```
CC DEMO.CC
```

When you load information from a disk, it is appended to the lines already in CC, so nothing is lost. Subroutines are ready to use when they have been read from a file, but new single-line function definitions and variable assignments read into the Input Window from a file are not executed until you move the cursor to them and execute them. For that reason, subroutines are displayed as ordinary text but new entries in the Input Window are highlighted.

See III.12.1 for information on saving a printable record of your entries to disk; see III.10.6 for information on saving a list to disk. For information on saving graphs to disk, see III.4.7 and III.4.8.

III.12.4 Controlling Output Precision

CC initially displays its results with 6 digits precision. You can change this with the command:

```
Precision(n)
```

where n is the number of digits precision you wish to display. Internally, all values are stored at full precision no matter what output precision you specify. The maximum precision in CC is 10 digits. The maximum precision in CC87 is 15 digits. To display all numbers in scientific notation, enter: `Precision(0)`.

III.12.5 Controlling Accuracy of Integration and Equation Solving

The operation `Solve(f(x),x=a)` uses Newton's method beginning with a first guess of $x = a$. It converges when two successive estimates *oldx* and x are found for the solution such that both $|f(x)|$ and $|oldx - x|$ are both less than:

$$\text{Min}\{1, |x|\}\textit{SolveTolerance}$$

(See V.3.2 for more details.) *SolveTolerance* is an internal constant in CC. Its default value is 0.0000001; you can change it to any value v with the command:

```
SolveTol(v)
```

This factor also controls the convergence of the operator `Implicit`. When the operator `FSolve` (fast solve) is invoked, *SolveTolerance* is temporarily set to 0.01. *SolveTolerance* is restored to its prior value at the end of this `FSolve`.

The convergence of the integration operator `In` is controlled by two constants. The operation `In(f(x),x=a to b)` first divides the interval (a,b) into three subintervals, and computes the midpoint of each. Based on these three subintervals, a midpoint approximation M and trapezoidal approximation T are computed for the integral. If T and M are approximately equal, then Simpson's approximation $(2M+T)/3$ is returned as the value of the integral; otherwise the process is repeated on each of the six subintervals of (a,b) that have been constructed. There are constants *IntegrationTolerance* and *IntegrationDepth* controlling this process. Integration converges on an interval if:

$$(b-a)\ \textit{IntegrationTolerance} < |T-M|$$

and the application of the algorithm to repeated subdivisions of the original interval will continue up to a depth of *IntegrationDepth*. That is, the maximum number of subintervals that can be constructed during the calculation of one integral is $6^{\textit{IntegrationDepth}}$. The default values are *IntegrationTolerance* = 0.0001 and *IntegrationDepth* = 5; these can be altered with the commands:

```
InTol(v)    // new value for IntegrationTolerance
InDepth(n)  // new value for IntegrationDepth
```

When `Fin` (fast integration) is performed, IntegrationTolerance is temporarily set to 0.1 and IntegrationDepth to 2. At the conclusion of the operation `Fin`, these constants are restored to their prior values.

III.12.6 Monitoring Memory Usage

When CC starts, it measures the memory available on your computer. The program halts immediately if insufficient memory is available. If only a small amount of memory is available for running CC, the program warns the user. While CC is running, the number that appears in the lower-right corner of the screen below the Output Window is the number of free bytes remaining in your computer's memory. A computer with the maximum 640K memory will start out with about 200,000 bytes free. This number changes dynamically while the calculator is calculating, so you can see how much memory is used by different calculator operations. CC frequently measures its free memory while calculating. If a complex operation threatens to overflow the available memory, CC will stop itself before crashing. CC also monitors the available stack space, and it will stop before crashing if stack space grows short. When CC encounters too little stack space or memory, it displays the error message: *Formula is too complex to evaluate--check for infinite recursion.*

The usual cause of running out of memory is infinite recursion, although extremely complex operations may exhaust your computer's resources. See V.2 for more information on recursion. Although CC measures its free memory frequently, it cannot do so constantly. It is possible but unlikely that CC will crash on an out-of-memory error. If it does crash for this reason, error numbers 202 or 203 will be displayed on the screen.

Every time CC draws a graph, it saves the graph so you can recall it by pressing function key F9. Normally CC saves the graph in memory, using a large block of available space. After you draw your first graph, you will see a precipitous decline in available memory. Saving a 3d-graph uses about 30,000 bytes; saving a 2d-graph takes about 30,000 bytes for a Hercules system, 128,000 bytes for an EGA system, and 150,000 bytes for a VGA system. If insufficient memory space is available to save the graph, then CC will redirect the graphic image to the current disk directory. If no disk is available, for example if you are running off floppy disks and have removed your program disk from the disk drive or if your disk is too full to hold the graphics image, then CC will warn you that the last graph was not saved. Press any key to remove the warning and continue. If you are running off floppy disks and need to save the last graphics image on disk, be sure your disk has enough space to hold the image and do not remove the disk after the image is saved if you plan to recall the image with the F9 key. A 720K or larger disk has plenty of room for the program CC.EXE, the help file CC.HLP, and the last graphics image. However, if you are using 360K disks (double-density 5" disks) and if you want to save the last graphics image on disk, you should prepare a disk with only the help file CC.HLP on it. After loading the program CC.EXE into your computer, replace the program disk with the disk containing only CC.HLP and run CC with this disk in place. Then CC's help feature (function key F1) will be available to you, and you will have room to save your last graph on disk.

You can force CC to save the last graphics image on disk instead of in memory by pressing Shift-F9. Shift-F9 is a toggle that switches CC between saving the last graphics images on disk or in memory. The current location for storing the last graphics image is always displayed below the available memory indicator. If there is a graphics image in memory when you press Shift-F9, then the image will be copied to disk and the available memory indicator will increase by an amount equal to the number of bytes necessary for saving a graphics image. WARNING: if for any reason the current graphics image in memory cannot be saved on disk, then the image will be lost. If the last graphics image is on disk, pressing Shift-F9 will switch it back to memory, and the available memory indicator will decrease accordingly. If insufficient memory is available for storing the graphics image, this command has no effect.

You can redirect the last graph file to any drive and directory you wish; see III.12.8.

Graphics images can also be saved in disk files for later recall by pressing F7 while the image is on the screen. This feature is useful for preparing illustrated talks with CC, since a graph can be recalled without the need for reconstructing it from formulas. See III.4.7.

If you find yourself running short of memory, the first thing you should do if possible is switch the last graphics image from memory to disk by pressing function key Shift-F9. Additionally, you can use the command `forget` to delete unnecessary functions and variables (III.3, III.9), and you can erase unused lines from the Input and Subroutine Windows. If you are running CC alongside other programs like print and disk buffers, Sidekick or Windows, you should save your work (F7), quit CC, disable these other programs, restart CC, reload your work (F8) and continue.

III.12.7 Shorthand Abbreviations and Greek Characters

You can use the following abbreviations for CC's most-used operators:

Abbreviation	*Meaning*
`GR`	`Graph`
`SK`	`Sketch`
`DTG`	`DotGraph`
`QU`	`QuickG`
`PO`	`PolarG`
`PA`	`ParamG`
`WI`	`Window`
`GR3`	`Graph3D`
`PA3`	`ParamG3D`
π = Alt-p	`Pi`
√ = Alt-q	`Sqrt`
Σ = Alt-s	`Sum`
`IMP`	`Implicit`
≤ = Alt-l	<=
≥ = Alt-m	>=

In addition, Greek letters can be used in CC's variable and function names, either alone or with other characters:

alt-a	α
alt-b	ß
alt-d	δ
alt-f	φ
alt-g	Γ
alt-h	ϵ
alt-o	Ω
alt-t	Θ

Remember, enter ê = 2.1728.. as `Alt-e`, î = $\sqrt{-1}$ as `Alt-i` and ² as alt-2.

III.12.8 Configuration File: Relocating Help File and Last Graph File

CC normally keeps its help file CC.HLP and its last graph file LASTGRAF.___ in the same directory as CC.EXE. (You can run CC from any directory. The path information you give DOS to start CC will enable CC to find its help file and last graph file.) You may want to relocate the help file and the last graph file to another directory, for example a ram disk or an unprotected directory on a network.

You may also want to save a color selection you have made with `TextColor`, so that CC begins with these colors when you start it. You can redirect the help file and last graph files and save color information by creating a configuration file CC.CFG with this information. While CC is running, press `Control-F9`. You will be offered the opportunity to enter a new path for your help file and last graph file, and--if you are running a color system--to save your current color scheme. The new paths to the help file and last graph file will take effect immediately, and they will be saved in a file CC.CFG in the same directory with CC.EXE. The next time you run CC from this directory, the changes recorded in CC.CFG will be effective.

You can change the configuration file at any time by pressing Control-F9 and entering new information. You can also delete or move the file and run CC without it to return CC to its default settings.

Your help file and any existing last graph file will not be moved when you create or change your configuration file. You must move the help file to the desired location before running CC and telling it about the new location for the help file, and any existing last graph file may be lost when you tell CC to use a new location for last graph files.

IV. PROGRAMMING WITH CC

CC includes a programming environment optimized for mathematical calculations. Although CC has enough commands to perform many calculations directly, CC can also be programmed to implement complex mathematical algorithms that would otherwise require a programming language like Pascal, FORTRAN or BASIC. In addition to implementing algorithms, CC's programs can display intermediate results or messages, accept input from the keyboard and display graphs. This chapter explains CC's programming capabilities. More examples of useful programs can be found in VI.2.

A program in CC is not a single mass of code as it would be in BASIC. Instead, you program CC by writing short subroutines, each with its own name, much as you would do in Forth or APL. The idea is not to write a single large program but to write a bunch of small subroutines that extend the power of CC. These subroutines can be invoked from the Input Window or by other subroutines just like CC's built-in functions and operators. Your subroutines will be written in a language similar to Pascal, with a few features borrowed from C and BASIC. Subroutines can be saved and loaded from disk files (III.12.2).

A caution about variables and user-defined functions: no variable or user-defined function from one subroutine can be referred to directly by another; not even global variables and user-defined functions from the Input Window are available to subroutines. CC's variables and user-defined functions are entirely untyped but totally local. To share information between operations, you must use the explicit parameter-passing mechanisms provided by CC's programming language. However, CC permits a broad range of data types to be passed as parameters to subroutines. Not only constants, but lists and even user-defined functions can be passed as parameters to subroutines, so you can write subroutines like integrating or equation solving algorithms that operate on arbitrary functions.

IV.1 How to Enter Subroutines

Subroutines are typed into their own window, called the Subroutine Window. You can view this window by pressing function key `F10`. `F10` toggles the screen between the Subroutine Window and the Input Window. Try pushing it a few times to see the screen switch windows. Entering subroutines into the Subroutine Window is almost the same as typing formulas into the Input Window. You can use the cursor keys to move around your text; the <Insert> key toggles between Insert and Overwrite mode; and the <delete> and <backspace> keys erase text. As in the Input Window, function key `F3` inserts blank lines into your text, and function key `Shift-F4` deletes lines. But the Subroutine Window differs from the Input Window in two ways. When you push <ENTER>, the line you are working on divides at the cursor, and lower lines are pushed down. Also, the cursor stops below the beginning of the prior line, not the left edge of the screen. In the Input Window, pushing <ENTER> executes a line without changing it and moves the cursor to the left edge of the screen below the executed line.

Lines in the Subroutine Window are never executed until the procedure or function they belong to is executed, but the lines are compiled as soon as you complete them. If you enter an incorrect line, a warning message is flashed on the screen. The contents of your Subroutine Window can be saved with your Input Window on disk. You can add previously saved subroutines to your current ones by loading them from disk (III.12.3).

Throughout this chapter, subroutines will be displayed in `10 point type` and commands for the Input Window will continue to be displayed in `12 point type`.

IV.2 Functions and Procedures

There are two types of subroutines: function subroutines and procedures. Each is a sequence of lines that is executed just like lines in the Input Window, but subroutine functions return a value while procedures do not. Here are two examples that illustrate the difference. You can try them out by switching to the subroutine window and typing them.

```
Procedure Greeting
    // Print 'Hello' on screen
  write('Hello')
end

Function Five
    // Return 5
  return(5)
end
```

Now push F10 to switch back to the Input Window, and enter the commands:

```
Greeting
a = Five
```

`Greeting` caused the message "Hello" to be flashed on the screen; `Five` returned the value 5. Each subroutine begins with a line containing the word `Procedure` or `Function` followed by the name of subroutine; and each subroutine ends with a line containing the single word `end`. In order for a subroutine function to return a value, the command:

```
Return(v)
```

must appear as one line of the subroutine. The term `v` can be a number or a numerically-valued expression or a list. The value returned by the subroutine function is determined by the last `Return` executed. If `Return` is not executed, then the subroutine function returns the value `UNDEFINED`. If an error occurs while a function subroutine is executing, even if it occurs after `Return` is executed, then the subroutine returns the value `UNDEFINED`.

Once defined, subroutines are invoked like built-in functions and operators by typing their names in the Input Window. Also, like functions and operators, subroutines can have arguments or parameters. We could write a subroutine to draw a circle of radius 1:

```
Procedure Circle
    // draw a circle of radius 1
  PolarG(1,t=0,2*pi)
end
```

However, it would be much more useful to create a subroutine that would draw a circle of radius *r*:

```
Procedure Circle(r)
    // draw a circle of radius r
  PolarG(r,t=0 to 2*pi)
end
```

To use the first procedure, you would just type the word `Circle` into the input window. To use the second procedure, you would have to specify the radius of the desired circle. The command `Circle(1.5)` would draw a circle of radius 1.5.

Subroutines can have more than one parameter. The next example draws a circle of radius *r* about the point (*a*, *b*):

```
Procedure Circle(r,a,b)
    // draw a circle of radius r and center (a,b)
  x(t) = a + r*cos(t)                 // (x(t), y(t)) is a point on the
  y(t) = b + r*sin(t)                 // circle at angle t from the
                                      // center (a,b)
  ParamG(x(t),y(t); t = 0 to 2*pi)    // graph the circle
end
```

To use this subroutine to draw a circle of radius 3 about the center (1,1), you would type:

```
Circle(3,1,1)
```

into the Input Window.

Subroutines can invoke one another. Unlike Pascal, a subroutine can even call another which appears after it in the Subroutine Window. Here is a subroutine that draws three concentric circles around the point (*a*, *b*) with radii *r*/3, *r*/2 and *r*.

```
Procedure ConcentricCircles(a,b,r)
    // draw circles with center
    // (a,b) and radii r/3, r/2, r
  Circle(r/3,a,b)
  Circle(r/2,a,b)
  Circle(r,a,b)
end
```

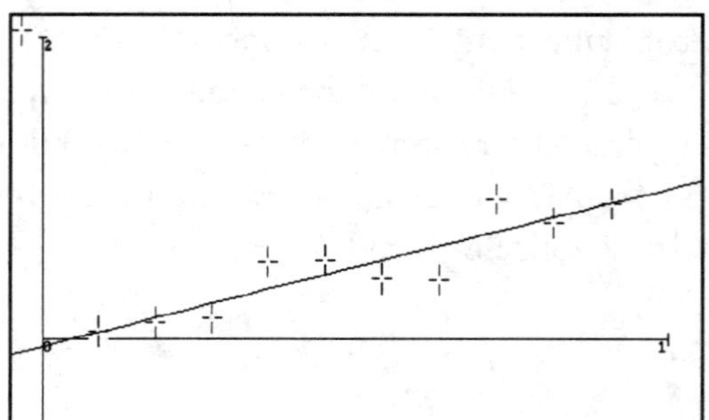

Fig. IV.1: Output from Procedure `Regression(10)`

The most powerful subroutines usually use the looping and alternation constructs presented below. However, we can present one more example using only the commands from Chapter III.

```
Procedure Regression(n)
    // produce n points near the line y = x, display them, and display
    // the regression line through them
  window(0,1,-1,2)
  a = list(j/(n+1),j=1,n)              // points on x-axis in interval
                                       // (0,1)
  b = list(a[j]+random/2-1/4,j=1,n)    // b[i] varies from a[i] by
                                       // random error from -1/4 to 1/4
  mark(a,b)                            // display points (a[i],b[i])
  r = regr(a,b)
  sketch(r[1]*x+r[2],x)                // regression line through the
                                       // points
end
```

Typical output from `Regression(10)` is shown in Fig. IV.1

IV.2.1 User-defined Function Names as Parameters

CC permits function names to be used as parameters for subroutines, so you can write a subroutine that applies to an arbitrary function. Simply put the sign % in front of the parameter name if you want it to refer to a function instead of a value. Of course, you must match the parameter to a function of the right number of variables (the number that the procedure expects to find in the function) when you call the procedure. Here's an example that approximates the derivative of an arbitrary function *f*(*x*) at any point *a* over a specified interval *eps*.

```
Function Der(%f,a,eps)
    // approximate the derivative of f(x) at x=a
  d = (f(a+eps/2) - f(a-eps/2))/eps
  return(d)
end
```

Note that the function name `f` was not preceded by `%` in the body of the procedure, only in the header.

To use `Der` to approximate $\left.\frac{d}{dx}Sin(x^2)\right|_{x=\sqrt{\pi}}$, enter:

```
g(x) = sin(x^2)
Der(g,sqrt(PI),0.001)
```

The function name in the call to the procedure is *not* preceded by `%`. Use `%` only in the procedure header to tell CC to expect a function name for this parameter; in the call to the procedure just put the function name without parentheses or `%`. The function name in the invocation of the procedure need not be the same name as the dummy function name in the procedure header.

A function parameter must be matched to a user-defined function name. It cannot be matched to the name of a subroutine or built-in function. You could not enter *Der(sin,pi,0.001)*.

IV.3 Program Control Structures

CC's programming language includes instructions for logical programming. You can create fixed or variable length loops and conditional alternatives in CC just as in any other high-level programming language.

The programming language also provides a special escape command:

```
exit
```

Anytime you execute this command, you will leave the current subroutine function or procedure, no matter how deeply imbedded you are in its control structures.

The control structures `For`, `If`, `Else`, `End`, `While` and `Repeat` introduced below can be used only in subroutines. If you try to use them in the Input Window, CC will display an error message.

IV.3.1 Comparison Operators

Program control is regulated by expressions that evaluate to the boolean values `TRUE` and `FALSE`. Such expressions are called boolean-valued expressions, and the simplest of them are expressions that compare two numbers. CC allows you to compare numbers six ways:

`a < b`	less than
`a <= b`	less than or equal to
`a > b`	greater than

`a >= b`	greater than or equal to
`a == b`	equal to
`a <> b`	not equals

CC uses the double equals sign == (like C) for "is equal to" to distinguish this operator from the assignment operator =. The reason for this departure from Pascal is to make it easier for non-programmers to write variable and function definitions without struggling with :=. (Actually CC will recognize := as equivalent to =, and it doesn't mind if you put semicolons at the ends of statements or append *then* to `if` and *do* to `while` or `for`. It just doesn't need these symbols.)

The result of a comparison is one of the values `TRUE`, `FALSE` or `UNDEFINED`. These values can be entered directly into CC too. Complex numbers and undefined values can be meaningfully compared with == and <>, but the other comparison operators always return `UNDEFINED` when applied to complex or undefined values.

In advanced programming applications, it may be necessary to identify whether or not a value is a list. You can do this with the operator

```
IsAList(a)
```

which returns true if `a` evaluates to a list and false otherwise.

IV.3.2 Logical Operators

You can combine boolean expressions in CC with the logical operators `and`, `or` and `not`. For example, to determine if the variable x is in the interval (a,b), use the expression:

```
x > a and x < b
```

In contrast to Pascal, logical operators in CC have a lower order of precedence than the comparisons, so the expression above works as expected.

The logical operators `and`, `or` and `not` can be used to combine boolean-valued expressions or to perform bitwise operations on integers (III.2.5), but integers and logical values cannot be mixed. In other words, `3 and true` equals `UNDEFINED`.

CC uses the "short-circuit" definition of `and` and `or`, as defined in the tables below. In the tables, *Numeric* includes all integer, hexadecimal and binary values including `UNDEFINED`, and *Symbolic* means any expression which includes undefined symbols.

a and b	b = True	b = False	b is Numeric	b is Symbolic
a = True	True	False	Undefined	b
a = False	False	False	False	False
a is Numeric	Undefined	Undefined	Numeric	a and b
a is Symbolic	a	false	a and b	a and b

a or b	b = True	b = False	b is Numeric	b is Symbolic
a = True	True	True	True	True
a = False	True	False	Undefined	b
a is Numeric	Undefined	Undefined	Numeric	a and b
a is Symbolic	True	a	a and b	a and b

Under short-circuit evaluation, the expressions *true or ...* and *false and ...* return *true* and *false* respectively, and the second term of the expression is not even evaluated. This is in line with contemporary programming practice and useful besides. While programming you will often encounter the situation where one condition must be true before another is meaningful. Typical is a subroutine entry like:

```
If IsAList(x) and x[1] == 0   //if x is a list with first entry 0
```

If x is not a list, then this condition should return false. It would be inconvenient if it returned `UNDEFINED` or an error message. Under short-circuit evaluation, the falsity of the first term will make the entire expression false and block evaluation of the second term. Were the second term evaluated, it would return an error message when x is not a list. The results for symbolic values are constructed on the assumption that symbolic values used in boolean expressions will eventually evaluate to the same type as the other term.

You can enter boolean expressions directly into the Input Window to see how they work. Here are some examples; for those labeled (short-circuit) only the first term is evaluated.

Formula	*Value*
`forget(x)`	so we can use x as an undefined symbol
`t = true`	convenient shorthand
`f = false`	
`t and f`	false
`f and t`	false (short-circuit)
`t or f`	true (short-circuit)
`f or t`	true
`not t`	false
`t or 3`	true (short-circuit)
`3 or t`	undefined
`t or x`	true (short-circuit)
`x or t`	x
`t and 3`	undefined
`f and 3`	false (short-circuit)
`3 and f`	undefined

IV.3.3 Conditional Statements

To write a useful program in CC or any other programming language, you need to be able to instruct the computer to make decisions based on the input data. For example, you might want to construct a function *sign*(x) that returns -1 when x is negative, 0 when x is 0, and 1 when x is positive. This function would be difficult (perhaps impossible) to define in the Input Window, but it can be defined as a subroutine function:

```
Function Sign(x)
    // returns -1 if x < 0; 1 if x > 0; and 0 if x = 0
  If x < 0
      return(-1)
    Else If x == 0
      return(0)
    Else
      return(1)
    end
end
```

The `If ... Else If ... End` construction is used to select from a list of options. The first option that evaluates to true is the one executed; the remainder are skipped. If you put an `Else` clause at the end of the list of options, then it will be executed if none of the other options is executed. The `Else` clause is not required; you can construct a sequence of options with the possibility that none are executed. You can have a single option too, a simple `If` clause with no `Else`'s.

CC uses the `If ... Else If ... End` construction much like other programming languages, but there are some restrictions you must be aware of. Each `If` and `Else If` must be followed by a boolean expression (an expression that evaluates to *true* or *false*) on the same line. Remember that a line in CC can occupy more than one screen line, but cannot be longer than 255 characters and must be typed without pressing `<ENTER>`. The commands to be executed should the `If` line be true must be the entries following the `If` line. There can be more than one such entry; they are terminated by the next `Else` clause or by the entry `end` signifying the end of the entire conditional statement. The final `end` is always required and must be on its own line.

Pascal programmers will have to alter some of their habits. You can add THEN after `If ...`, but you do not have to. CC terminates the boolean condition controlling `If` by the end of the entry. BEGIN is never used in CC, so in particular multi-line clauses following `If` do not start with BEGIN. Here are two examples of `If` constructs, showing how CC differs from Pascal:

CC

```
If x == 0
    y = 1
    z = 2
  Else If x < 0
    y = 2
    z = 1
  Else
   y = 0
   z = 0
  End
```

PASCAL

```
IF x = 0
  THEN
   BEGIN
      y := 1;
      z := 2
   END
     ELSE IF x < 0 THEN
    BEGIN
      y := 2;
      z := 1
        END
     ELSE
       BEGIN
          y := 0;
          z := 0
       END;
```

CC	*PASCAL*

`If x < 3`
`   y = 2`
`  end`
	`IF x < 3 THEN y = 2;`

IV.3.4 For Loops

You can use two kinds of loops in CC, fixed length loops controlled by `For` and variable-length loops controlled by `While` and `Repeat`. We begin with fixed-length loops.

The construct:

```
for x = a to b step c
    .....
  end
```

causes CC to execute the lines between the `For` clause and `End` multiple times, the first time with the variable `x` = `a`, the second time with `x` = `a+c`, and so forth until the last time `x` is not larger than `b`. The phrase `step c` can be omitted; in that case the step-value defaults to 1. This repetition is similar to the repetition in the `List` and `Sum` operators (III.9, III.10). Multiple lines can be included in a `For` statement, but whether one or more lines is used, the statement must be terminated by `End` on its own line. Here is a sample subroutine function that uses a `For` loop. It computes the midpoint approximation to the integral of an arbitrary function $f(x)$ on the interval (a, b) using n subintervals.

```
Function Mid(%f,a,b,n)
    // compute the midpoint approximation to the integral of f(x) on
    //the interval [a,b] using n subintervals
  dx = (b-a)/n                        // width of each subinterval
  M = 0;                              // accumulate sum of heights
  For x = a+dx/2 to b-dx/2 step dx    // For every subinterval
      M = M + f(x)                    // add another midpoint value
    End                               // End for
  Return(M*dx)                        // sum of midpoint values * width
End                                   // Mid
```

Here is another example subroutine using For loops. This one sorts the elements of a list *a* into increasing order.

```
Function Sort(a)
    // a is a list.  Returns a sorted copy of a
  for j = 1 to length(a)-1              // put smallest element from
                                        //  a[j..n] into a[j]
      for k = j+1 to length(a)          // For all elements a[k] after a[j]
          If a[j] > a[k]                // if a[j], a[k] out of order
              x = a[j]                  // swap them
              a[j] = a[k]
              a[k] = x
          end
      end                               // Now a[j] =min(a[j..n])
  end                                   // Now a[1] ≤ a[2] ≤ ... ≤ a[n]
  return(a)                             // return sorted copy of a
end
```

Note that invoking this procedure from the Calculator with the command:

```
y = Sort(x)
```

puts a sorted copy of x into y; the variable x is unaffected. To sort a list itself, you must use the command:

```
x = Sort(x)
```

IV.3.5 While and Repeat Loops

Loops controlled by For are useful in programming, but they are inflexible in that they require you to specify ahead of time how many times they are to be repeated. CC also includes loops controlled by While and Repeat which execute until some specified condition changes. The general forms of While and Repeat loops are as follows:

```
while <boolean expression>
    .....
  end
```

```
repeat
    .....
  until <boolean expression>
```

When CC encounters a `While` clause, it evaluates the associated `<boolean expression>`. If the expression is false, CC jumps to the entry following `End`; otherwise CC executes the entries following `While` until it comes to `End`. Then it goes back to `While` and re-evaluates the `<boolean expression>`. If the expression is true, the lines following `While` are executed again, otherwise CC skips to the line following `End`. This process is repeated until the `<boolean expression>` becomes false. Obviously, the expression must contain a variable that is altered within the `While` statement. Otherwise you would never exit the loop.

The `Repeat` statement is much like the `While` statement, except that the controlling condition is first evaluated after the body of the loop is executed once. At that point CC keeps executing the body of the loop until the condition becomes true.

As with other control statements, the term `While`, `Repeat`, `End` and `Until` together with their associated `<boolean expression>`'s must be on their own lines.

Here is a subroutine function using a `While` loop. It simulates a *geometric process*, which is an event that has a probability p of happening in each time interval. The function returns the number of time intervals before the event takes place. Since the function is based on the random number generator, it can return a different value each time it is called.

```
Function Wait(p)
    // assume an event has probability p of occurring in each time
     // interval beginning with time interval 0.  Return number of time
     // intervals until event takes place.  This function may return a
     // different value each time it is executed.  It models a binomial
     // process.
  n = 0                   // number of times event fails
  while Random > p        // Each time event fails
      n = n+1             // add one to counter
    end
  return(n)               // number of times event failed
end
```

Our next example uses a `Repeat` loop. It produces a list of integers representing a random walk with step size 1 that starts at 0 and continues until it reaches 5 or -5.

```
Function RandomWalk
   // Return a list of integers representing a random walk beginning
    // at 0 and ending at 5 or -5
  w = () ¦ 0                    // list containing the random walk
  last = 0                      // last position in the walk
  repeat
     Last = last + 2*Random(2)-3   // take a step +1 or -1
     w = w ¦ last               // append next step to walk
    until abs(last) = 5         // until walk reaches 5 or -5
  return(w)                     // list of steps in walk
end
```

The syntax of `While` and `Repeat` in CC is similar to the syntax of Pascal, with the following differences:

- DO is not required (but permitted) after `While` and `For`
- `While`, `Repeat`, `Until` and `End` must be on their own lines
- BEGIN is neither required nor permitted
- `End` is required to terminate a while loop, even for one-line loops

IV.4 Special Programming Operations

CC includes special facilities for adding interactive and graphical features to your subroutines. Some of these operations can be used only in subroutines and not in the Input Window.

IV.4.1 Strings

Alphanumeric strings are legitimate values for the variables in CC. You assign these values with commands like:

```
a = 'abc'
```

using apostrophes to delimit the string. This works both in subroutines and in the Input Window. String values can be compared with all the comparison operators, and the comparisons are case-sensitive. Here are some examples:

```
a = 'abc'
b = a
c = 'ABC'
a == b        // true
a == c        // false
a < b         // false
c < a         // true--upper-case letters precede lower-case
```

There are no special string-manipulation operators in CC.

You can use an apostrophe as a character within a string by doubling it. The string *x' = 2* is written in CC as `'x'' = 2'`.

Be careful not to confuse undefined variable names with strings. If you make the following two assignments:

```
a = 'ABC'
b = ABC
```

then the value of a is a string, while the value of b is an undefined variable. If you then make the assignment `ABC = 3`, then the value of b becomes 3 but the value of a remains the string `'ABC'`.

IV.4.2 Writing Output

To display a message in a box superimposed on the screen use the command:

```
Write(a,b,c,..)
```

The entries `a, b, ...` can be any legal CC expressions. They must be separated by commas. For example, if x equals 5, the command

```
Write('The first result is',x^2-2)
```

displays the message:

```
The first result is 23
```

The `Write` statement can be used to write results or instructions to a user. It can be used in both text and graphics mode (see IV.4.4 below for graphics mode). It is also useful for adding tracers to a subroutine that is not performing as expected.

IV.4.3 Obtaining User Input

CC enables you to input data while a program is running in either text or graphics mode (see IV.4.4 below for graphics mode). You are limited to inputting a single numerical value, but the value can be entered as a formula, not just a number. To make CC pause while running a subroutine to allow you to enter data, use the command:

```
Input('<message>',<variable>)
```

When this line is executed, CC stops, displays the string <message> in a box on the screen, and waits for you to type a number or formula and push `<ENTER>`. The value entered by the user is placed in <variable>. A sample input statement could be:

```
Input('Enter lower value', low)
```

You could respond `3` or `Sin(5)` or `2/3`, etc. Any legitimate CC formula using only constants and built-in functions is valid. The message can be omitted, and CC will substitute a generic message asking the user to input a value.

CC expects a numerical value in response to `Input`. If you want to have CC pause and accept a string value from the keyboard, for instance if you want the user to respond 'Y' or 'N' to some query, then you must use the statement:

```
InputS('<message>',<variable>)
```

This statement causes CC to stop, display <message> in a box on the screen, and wait for you to type something and press `<ENTER>`. Whatever you type is stored in the variable. The input is not evaluated. You might use `InputS` to control a loop like this:

```
repeat
    ...    // do something useful here
    InputS('Do you want to continue (Y/N)?',response)
  until response == 'n' or response == 'N'
```

This loop will be repeated until the user enters 'n' or 'N' in response to the query in the `InputS` statement.

IV.4.4 Graphics Programming

All the graphics commands in CC (III.4) can be included in subroutines, but normally after each one CC will pause and wait for you to press `<Enter>` or `<Esc>`. Sometimes you will want a program to draw several graphs before stopping, or you will want to build a picture out of points and line segments. CC has a command that keeps it in graphics mode while it executes a series of graphics commands:

```
Graphics
```

Any graphics commands following `Graphics` will affect the screen, but CC will not pause after each one. You turn off graphics mode with the command:

```
Text
```

When this command is issued, the computer pauses, displays the crosshairs, and waits for the user to push `<Enter>`. Both of these commands can be used only in subroutines.

Three-dimensional graphics commands and `Histogram` may not be used in graphics mode, although they can be used in a subroutine. Since the output from these commands cannot be combined with the output from other graphics commands, there would be no point in putting CC into graphics mode before executing them.

Although you may not want to return to text mode after each graphics command, you may want to pause in the midst of a sequence of graphics operations to view the screen before adding to it. While in graphics mode, you can cause the computer to pause between commands and wait for the user to push `<Enter>` or `<Esc>` by executing the command:

```
Pause
```

When your subroutine is paused, the user can reposition the crosshairs before pushing `<Enter>`; see VI.2.6 for an example of pausing a routine to permit the user to select a spot on the screen before continuing.

While in graphics mode you can clear the screen with `Erase` or clear and resize the graphics window with `Window` (III.4). The commands `Plot`, `Line`, `Rect`, `Mark`, and `UnMark` can all be used in graphics mode to construct diagrams.

For some graphics operations, it is necessary to determine the boundaries of the graphics window. For example, you might wish to stop drawing a level curve in a vector field when it ran out of the window. There are four functions that return the window boundaries:

```
windowleft
windowright
windowtop
windowbottom
```

The following user-defined function, which could be added to a subroutine or entered into the Input Window, returns true if the point (x, y) is in the current graphing window and false otherwise.

```
OnScreen(x,y)  = x >= windowleft and x <= windowright
            and y >= windowbottom and y <= windowtop
```

For our first example of a graphics subroutine, we present a procedure that places n bugs on the circumference of a circle and has each chase its clockwise neighbor until all end up in the middle of the circle (Fig. IV.2). Their paths are called pursuit curves.

```
Function dd(a,b,c,d)
    // distance (a,b) to (c,d)
  return(sqrt((c-a)²+(d-b)²))
end

Procedure Chase(n)
    // Illustrate pursuit curves with n points
  Window(-1.4,1.4,-1,1)            // Set viewing window
  Graphics                         // Turn on graphics mode
  LineStyle(0)                     // Use solid lines
  KeepLine
  PolarG(1,t=0,2*pi,200)           // Draw a circle of radius 1
  x = list(cos(j*2*pi/n),j=1,n)    // x-coordinates for bugs
```

```
    y = list(sin(j*2*pi/n),j=1,n)  // y-coordinates for bugs
    d1 = dd(x[1],y[1],x[2],y[2])   // distance from bug[1] to bug[2]
    while d1 > 0.03 do             // distance between bugs > 0.03
        a = x[n]                   // remember n'th bug
        b = y[n]
        for j = 1 to n do          // for each bug
            c = x[j]               // (c,e) is j'th
            e = y[j]               // (a,b) is j-1'st
            f = c + (a-c)/d1*0.03  // Move j'th bug toward j-1'st
            g = e + (b-e)/d1*0.03
            line(c,e,f,g)          // draw line old position to new
            x[j] = f               // update data for j'th bug
            y[j] = g
            a = c                  // advance (a,b) to j'th bug
            b = e
          end
       d1 = dd(x[1],c,y[1],e)      // update distance between bugs
     end
  end
```

The variables a,b,c,e,f,g in this procedure are not strictly necessary. They could be replaced by repeated references to x[j], y[j], etc., but they speed up the algorithm. CC uses linked lists to store lists, which is why you can use lists without first dimensioning them. But linked lists access their elements slowly. Each time you refer to x[j], the entire list x up the *j*'th element must be traversed. Thus it pays to use auxiliary variables in place of repeated references to list entries.

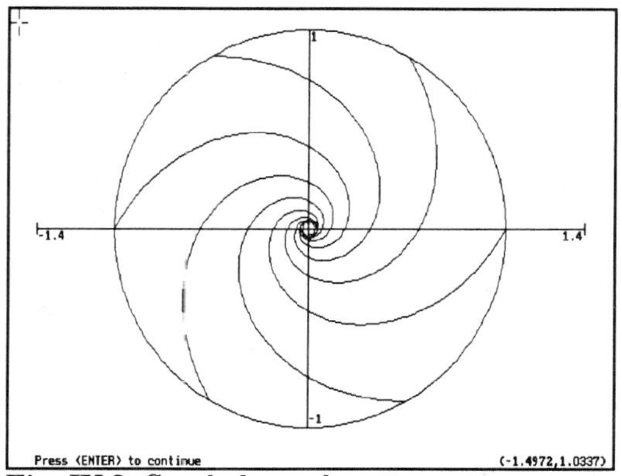

Fig. IV.2: Graph drawn by Chase(6)

As our final example of graphics programming, we present a subroutine that places n points symmetrically on a circle and connects them all by lines. The resulting pattern for n = 13 can be seen in Fig. IV.3

```
Procedure Design(n)
    // Place n points symmetrically on a circle and connect them all
    // to each other
  window(-1.4,1.4,-1,1)         // set viewing window
  graphics                      // turn on graphics mode
  PolarG(1,t = 0 to 2*pi)       // draw a circle radius 1
  x=list(cos(j*2*pi),j=1 to n)  // x and y coordinates of n points
  y=list(sin(j*2*pi),j=1 to n)  // placed on the circle
  for j = 1 to n                // for each point
     for k = j+1 to n           // for all points beyond it
          Line(x[j],y[j],x[k],y[k])    // connect with a line
       end
    end
  text                          // return to text mode
end
```

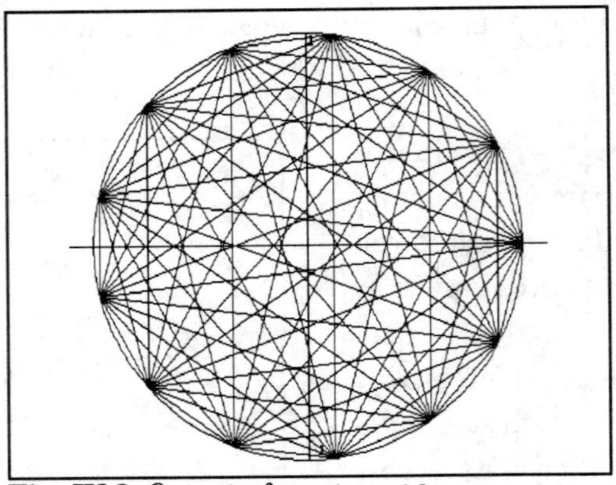

Fig. IV.3: Output of `Design(13)`

IV.4.5 Sound

You can make the computer emit a short, not unpleasing sound by entering the command:

```
beep
```

This command can be used in programs to indicate that input is required or an error has occurred. It is also useful in debugging subroutines to signal that some part of the code has been executed.

V. TECHNICAL NOTES

This chapter discusses three issues related to the way CC works internally. You can use CC without reading this chapter, but it will help you use CC better. The concepts presented here are especially important for programming CC.

V.1 Variables: Free, Bound and Local

CC uses three kinds of variables. When you assign a value to a symbol, or when you use a symbol in a formula that is not a bound or local variable, you are using a *free variable*. A free variable is one whose value is determined outside the formula that uses it. Free variables are listed in the output window. When you use an undefined symbol in a formula, for example you write `x+2` with `x` undefined, then CC treats `x` as a free variable whose value is its own name. A free variable cannot have the same name as a function, and there can only be one free variable with a given name.

A *bound variable* is a variable used in a function definition. When you write

```
f(x) = x^2-x+1
```

the variable `x` is a bound variable. Bound variables are just place holders; when you enter:

```
f(3*a-2)
```

the expression `3*a-2` is evaluated, and its value is substituted for each instance of the bound variable `x` in the definition of `f(x)`. The name used for a bound variable is local to the expression in which it is used. That name can be the same as the name of any other variable or function in CC, and there will be no confusion. CC knows when a variable in the expression defining a function is a bound variable and when it is a free variable. The only restriction is that you cannot refer to a free variable with the same name as the bound variable in a function definition. If you had a free variable called x, and you defined `f(x) = x+x^2`, you could not have the first x refer to the bound variable and the second x refer to the free variable; both x's refer to the bound variable.

Finally, CC uses *local variables* for its operators. When you write:

```
In(Sin(x),x=0 to pi)
```

then x is a local variable. Like bound variables, local variables can use the same name as a free variable without confusion, because they cannot be used except inside the operator where they are defined. Local variables are defined only for the expression to their left within the operator using them. The expression

```
In(Sin(x), x=0 to x)
```

means exactly the same thing as:

```
In(Sin(t), t=0 to x)
```

In the first version, the local variable x is the variable of integration for the function `Sin(x)`; in the expressions defining the limits of integration, x has no special meaning unless it is defined as a free variable. Since the local variable would hide a free variable of the same name in the integrand, the two expressions:

```
In(Sin(x*x), x=0 to pi)
In(Sin(x*t), t=0 to pi)
```

would not mean the same thing. The first means the same as $\int_0^\pi \mathrm{sin}(x^2)\,dx$,while the second means $\int_0^\pi \mathrm{sin}(xt)\,dt$. In CC's integration, the local variable takes the place of the differential symbol.

CC operations that use local variables are integration, differentiation, equation solving, graphing, sums, products and list construction. In the expression:

```
f(x) = In(a*x*y,y=a to x)
```

a is a free variable, *x* a bound variable and *y* a local variable.

The use of variables in differentiation is complicated by the option of omitting the last term. If you differentiate:

```
dif(f(x), x=a)
```

the expression `f(x)`, which includes the local variable x, is formally differentiated with respect to x, then the value a is substituted for x. If you omit the final =a in differentiation, then CC inserts the expression =x where x is the name of the local variable. That is, the following expressions are identical for CC:

```
dif(f(x), x=x)
dif(f(x), x)
```

But that second x is not the local variable for `dif`; it could be a bound variable if the derivative appears on the right of a function definition using x as a bound variable or else it could be a free variable. Thus:

Short Version	*Long Version*	*Value*
`x = 3`	`x = 3`	3
`dif(x^2,x)`	`dif(t^2, t=x)`	6
`f(x) = dif(x^2,x)`	`f(x) = dif(t^2,t=x)`	x^2

Note that in the third line the bound variable x takes precedence over the free variable of the same name.

Free variables and user-defined function definitions from the Input Window cannot be referenced by subroutines, and free variables and user-defined functions defined in one subroutine cannot be used in another. The only globally defined objects in CC are subroutines.

V.2 Recursion

Recursion occurs when a definition refers, directly or indirectly, to itself. On first glance, recursive definitions often appear to be circular, but close examination will reveal that correct recursive definitions do not create a vicious circle. CC defines three kinds of objects:

- subroutine functions and procedures
- free variables

- user-defined functions

Each can be used recursively. Here is an example of a recursive subroutine function:

```
Function Fibonacci(n)                    // Returns n'th Fibonacci number.
  If n <= 1                              // If n <= 1,
      Return(1)                          // n'th number = 1
    else                                 // otherwise
      Return(Fibonacci(n-1)+Fibonacci(n-2)
                                         // sum of previous two numbers
    end
end
```

A recursive variable definition is an entry like:

```
a = a + 2
```

and a recursive user-defined function definition is one like:

```
f(x) = dif(f(x+2), x)
```

In each of these examples, the object being defined--the subroutine function `Fibonacci(n)`, the free variable a and the user-defined function `f(x)`--appears in the definition. These are all examples of direct recursion. Indirect recursion occurs when an object is defined in terms of other objects whose definitions refer to the first object, for example:

```
a = b + 2
b = a - 3
```

Recursion is logically legitimate provided the backwards chain of references eventually stops. In the subroutine `Fibonacci(n)` above, the backwards self-references eventually stop. Evaluating `Fibonacci(1)` or `Fibonacci(0)` will return the value 1; no recursion is involved. Evaluating `Fibonacci(3)` does use recursion. `Fibonacci(3)` invokes `Fibonacci(2)` and `Fibonacci(1)`; and `Fibonacci(2)` invokes `Fibonacci(1)` and `Fibonacci(0)`. Thus evaluating `Fibonacci(3)` results in four more calls to the subroutine function `Fibonacci`, but eventually `Fibonacci(3)` returns a value of 3.

Infinite recursion occurs when a chain of backward references is created without any terminus. The subroutine procedure `R(n)` in the next section illustrates infinite recursion. Definitions using infinite recursion are genuinely and viciously circular.

V.2.1 Recursion in Subroutines

Recursion with subroutine functions and procedures is the easiest recursion to understand. Subroutines are stored as lists of lines just as you enter them. The individual lines are compiled as they are entered (so an error message will flash on the screen if you enter an erroneous line), and the subroutines are executed line by line when they are called from the Input Window or by another subroutine. Of course, the control structures can alter the order in which the lines are executed.

Each invocation of a subroutine function or procedure has its own set of variables. That is, if x is a variable in a subroutine, a new memory location is reserved for x each time the subroutine is invoked. The memory location is returned to the heap when the subroutine terminates. If the subroutine invokes another while it is running, then the memory locations for the first are retained while new memory locations for the second are created. If x is also the name of a variable in the second subroutine, then a new memory location is reserved for this new x, so values placed in x by the second subroutine do not destroy values left in x by the first subroutine.

Now suppose that the first and second subroutines are the same, or that a subroutine calls itself. A new set of memory locations for the subroutine's variables would still be created for each invocation of the subroutine. The value of x in one invocation of the subroutine would be unaffected by changes made to x in another invocation. However, if a subroutine was caught in an infinitely recursive loop, each call to the subroutine would reserve additional memory until your computer had no more to give. What would happen then?

CC monitors its own heap (III.12.5), and it will not crash on infinite recursion. Instead, it stops itself before running out of memory and displays an error message. To see the error message generated by recursion calls that go too deep, and to find out how deep recursion calls can go on your computer, enter the following procedure into the Subroutine Window and invoke it from the Input Window with the command `R(1)`:

```
Procedure R(n)
    // Write n, increment n, and do it again
  Write('Recursion depth',n)
  R(n+1)
end
```

This procedure illustrates infinite recursion because it will apparently create a chain of calls to itself without end. A computer with 640K memory and Hercules-style video achieves a depth of 31 before this recursion is stopped for lack of memory space.

V.2.2 *Recursion in Variables*

Variables are assigned values when their definitions are executed. Thus, if you define a variable:

```
b = a+2
```

the value that is stored in b will depend on the value in a when the definition is executed. If a is later given a different value, the value of b will not change. If a was undefined when this line was executed, then b would have as its value the algebraic expression $a+2$. See III.3 for more details on this point.

Once a is defined, you can execute the recursive command:

```
a = a+2
```

which increments the value of a by 2. However, if a is undefined when this line is executed, then an anomolous situation can develop.

The initial effect of executing this line is to store the expression $a+2$ in the variable a. Then any future reference to a will result in an infinite recursion by causing CC to evalute the expressions a and 2 prior to summing them. But evaluating a will again cause CC to evaluate a and 2 prior to summing them. This process continues until CC runs out of heap or stack space and displays the error message for infinite recursion.

V.2.3 Recursive User-defined Function Definitions

When you define a function, for example:

```
g(x) = sin(x+a)
```

the expression on the right of = is stored in the computer's memory. In particular, the constant a is stored as a literal symbol; it is not evaluated when the definition for $g(x)$ is stored. When you later evaluate the function $g(x)$ by entering, for example, `g(2)`, the value you get depends on the value of a at the time you perform the evaluation.

Because no component of a function is evaluated until the function is invoked, it is not obvious how to define functions recursively. The second line in the sequence of entries:

```
g(x) = sin(x)
g(x) = dif(g(x), x)
```

attempts to define `g(x)` recursively. But the result is not $g(x) = cos(x)$. Instead, the second definition simply replaces the first, and the replacement is circular. The first time the replacement `g(x)` is used, for example attempting to compute `g(pi)`, will result in infinite recursion.

Nevertheless, it is occasionally useful to define functions recursively. You can do this with the command `Eval`, which is used to force a function definition to evaluate its defining terms immediately. If we slightly alter the commands above:

```
g(x) = sin(x)
g(x) = Eval(dif(g(x), x))
```

their effect will be completely different. Instead of having the circular definition $g(x) = g'(x)$, we have the useful definition $g(x) = \cos(x)$. The command `Eval` forced CC to evaluate the term to the right of the equals sign when $g(x)$ was defined for the second time, and CC used the original definition of $g(x)$ to create the second definition.

The `Eval` command is particularly useful for taking repeated derivatives of a function. You can define the following subroutine function in the Subroutine Window to compute the value of the n'th degree Taylor polynomial about 0 for the function $f(x)$ at $x = a$:

```
Function Taylor(%f,n,a)
    // returns the value of the n'th degree Taylor polynomial of f(x)
    // about 0 at x = a
  T = f(0)                         // Constant term of Taylor polynomial
  For i = 1 to n
      f(x) = Eval(dif(f(x),x))     // i'th derivative of original f(x)
      T = T + f(0)*a^i/i!          // add f(i)(0)ai/i! to Taylor polynomial
    end
  Return(T)
end
```

Note that this version of the Taylor polynomial requires fewer computations than that found in III.11.2, since the derivative in each term uses the already-computed derivative from the previous term. It could be made even more efficient by replacing the two innermost lines with:

```
f(x) = Eval(a*dif(f(x),x)/i)        // aif(i)(x)/i!
T = T + f(0)
```

When you apply `Eval` to an expression, two kinds of components are not replaced by their definitions: calls to `Random` and calls to subroutines. You would not want to replace calls to `Random` with fixed random values when you used `Eval` to define a function, and it is not possible to replace a call to a subroutine with the entire subroutine inside a one-line user-defined function.

V.3 Remarks on the Algorithms Used in CC

Before using any computer program for critical calculations, you should assure yourself that the program behaves as you expect. You should try some similar calculations with known results to see that the program returns the correct values. You should exercise the program extensively to determine if it contains any quirks or bugs. And you should understand the algorithms used by the program to know how the program arrives at the answers it is giving you. This last step is particularly important for determining the accuracy of the answers you are getting. The following sections discuss the algorithms that CC uses to evaluate integrals, solve systems of equations, and compute the gamma function. For computing other transcendental functions including logarithms, exponentials and trigonometric functions, CC uses the algorithms provided by Borland's *Turbo Pascal®* extended by the usual arithmetic formulas to complex values.

CC stores real values in Turbo's Real format, which stores each number in exponential form using a 39 bit mantissa and 8 bit exponent. That means that CC stores numbers with about 12 digits accuracy and a maximum exponent of 36. (CC-87, the version of CC that runs with the math coprocessor, stores numbers in Extended form using a 63 bit mantissa and 15 bit exponent. That means that CC87 stores numbers with about 19 digits accuracy and a maximum exponent of 4899.) Integer values up to 32 bits are stored with complete accuracy. Turbo's algorithms are generally quite good, giving results accurate to the full precision of the numbers stored. However, trigonometric functions (including logarithms and exponentials applied to complex arguments) are not so accurate when applied to large values. CC restricts trigonometric functions to real arguments of absolute value less than 10^9 (10^{15} for CC87). For arguments this large, Turbo's trigonometric functions are accurate to only four digits. It seemed pointless to permit even larger arguments and correspondingly less accurate results.

V.3.1 Integration

CC's adaptive Simpson's algorithm for integration is described in III.12.4; what is not discussed there is CC's stochastic modification of this common algorithm. Some numerical analysis programs fail when trying to integrate a periodic function with short period. They sample the function at a relatively small number of points which are distributed uniformly on the interval of integration and they happen to select points where the function is zero. They then conclude that the integral must be zero. To test a program for this error, try the following integral:

$$\int_0^{2\pi} \sin^2(196\,t)\,dt$$

When CC divides an interval into three parts to begin computing its approximation to an integral, it does not divide the interval into equal parts. Instead, it modifies the two internal points of the equal tripartite subdivision by a random amount of up to 10%. This prevents CC from misinterpreting a highly periodic function as a constant function. The midpoints of these three subdivisions are the actual midpoints, so CC's Simpson's approximation is still an approximation to the integrand by three parabolas, one on each of the three subdivisions made to the integral. However, every time you evaluate an integral with CC, you may get a slightly different answer. To see this, enter the command `Precision(10)` before computing an integral twice, and compare your answers. (If you are using CC87, try `Precision(15)`.)

Another way of investigating the stochastic nature of CC's integration algorithm is to use the command `InteG`. Try this:

```
Window(0,1,0,1)
InteG(x,x=0,1)
```

If you execute the `InteG` command twice, you will see that the vertical lines it draws do not fall in exactly the same place each time. The lines are drawn at the points where the function is sampled for computing the integral.

V.3.2 Equation Solving

CC uses Newton's method to solve systems of equations. Given a system of equations:

$$\begin{array}{ccc} f_1(x_1, \dots, x_n) & = & g_1(x_1, \dots, x_n) \\ \vdots & \vdots & \vdots \\ f_n(x_1, \dots, x_n) & = & g_n(x_1, \dots, x_n) \end{array}$$

and initial guesses:

$$x_1 = a_1 \quad \dots \quad x_n = a_n$$

CC forms the Jacobean matrix $J = [\partial(f_i - g_i)/\partial x_j]$ of the functions $f_i - g_i$, a scalar vector $\mathbf{x} = (a_1, \dots, a_n)$ and a vector of functions $\mathbf{f} = (f_1 - g_1, \dots, f_n - g_n)$. Then CC forms a better approximation $\mathbf{x}$ to the solution of the system by iteratively applying the formula:

$$\begin{aligned} \mathbf{dx} &= J(\mathbf{x})^{-1}\mathbf{f}(\mathbf{x}) \\ \mathbf{x} &= \mathbf{x} - \mathbf{dx} \end{aligned}$$

The solution algorithm uses a constant *SolveTolerance*, which CC initializes to 0.0000001 and which can be reset by the command `SolveTol(x)` (see III.12.4). Using *SolveTolerance*, CC computes an integer constant

$$MaxCount = \text{Round}(5\text{-}(25/6)(\ln(\text{SolveTolerance})+1))$$

Thus *MaxCount* is logarithmically linear with *SolveTolerance*; *MaxCount* = 5 when *SolveTolerance* = 0.1; and *MaxCount* = 30 when *SolveTolerance* = 0.0000001, its default value.

CC stops iterating the solution algorithm when two conditions are satisfied:

$$|\mathbf{f}(\mathbf{x})| < \max(\mathbf{x}, 1)\textit{SolveTolerance}$$
$$|\text{d}\mathbf{x}| < \max(\mathbf{x}, 1)\textit{SolveTolerance}$$

or *MaxCount* iterations have been made. The norm used is the absolute-value norm, not the Euclidean norm, in case $n > 1$. The reason for the first stopping condition is clear; it guarantees that **f**(**x**) is nearly 0. The second condition prevents the algorithm from stopping on a pretty good solution when it might go on to a better one. It is necessary for solving functions whose graphs are tangent or nearly tangent to the x-axis.

If the solving algorithm stops because of excessive iterations, it accepts the final value of **x** as an adequate answer if:

$$|\mathbf{f}(\mathbf{x})| < 100 \max(\mathbf{x},1)\textit{SolveTolerance}$$

That is, CC tries for a very good answer but, in the end, will accept a pretty good one. If this last condition is unsatisfied, then an error message is returned.

Anytime you are not certain of the result returned by `Solve`, you can check the result by substituting it back into the original equations.

V.3.3 Factorials and the Gamma Function

When starting up, CC computes a table of factorials. Thus CC's factorial function is a fast table lookup, not repeated multiplication. CC's algorithm for the LnGamma function is taken from Press, Flannery, Teukolsky and Vetterling; *Numerical Recipes in Pascal*. Cambridge University Press, 1989.

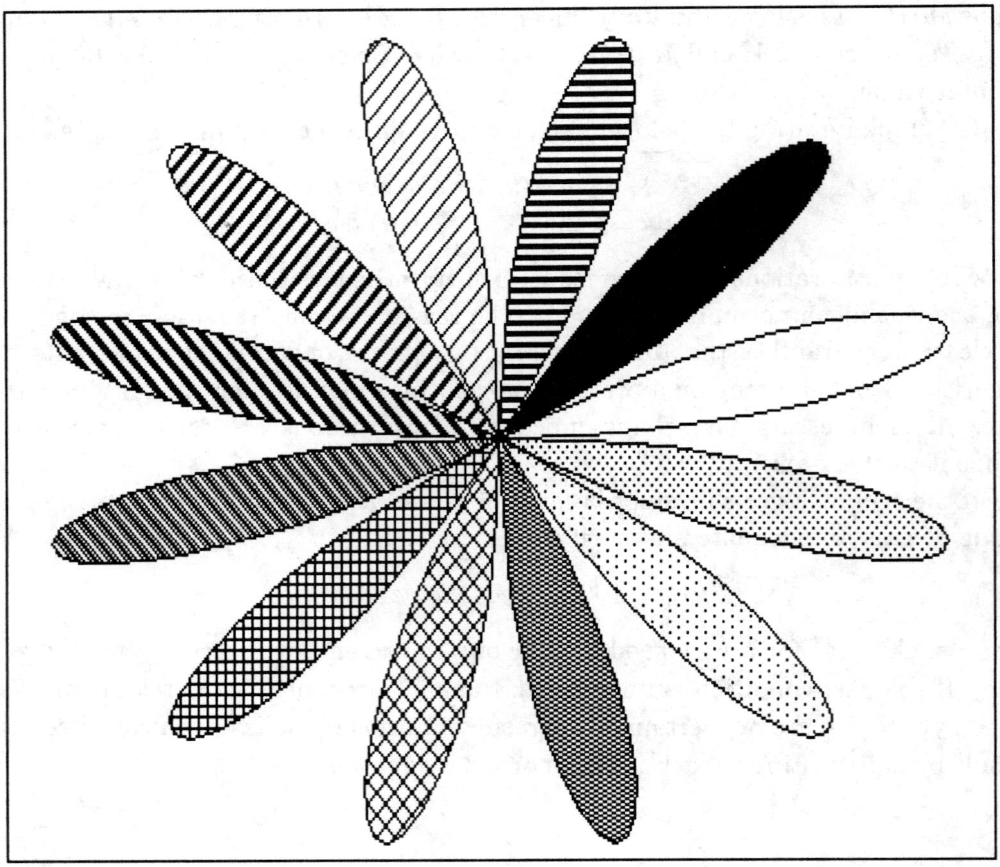

VI. APPLICATIONS

This chapter is divided into two parts. In the first part we present a number of non-programming applications, mostly derived from calculus problems. The second part contains programming examples including advanced techniques of graphing, numerical analysis and chaos theory. Each example contains all of the CC commands necessary for solving the problem.

VI.1 Calculator Applications

The applications which follow demonstrate how CC can be used to solve problems at the level of calculus and advanced calculus. Most are approached in three steps: define the necessary functions; make a graph illustrating the problem; then compute the final answer. The graphing capabilities of CC encourage problem solving by exploration, and this method is illustrated in the solutions below.

VI.1.1 Integration and Equation Solving

***Problem*:** Find the area bounded by the x-axis and the curve $y = \sin(x) - x^2$.

***Solution*:** First define the desired function and graph it:

```
f(x) = sin(x) - x^2
Window(-1,1,-1,1)
quickg(f(x),x)
```

The graph is shown in Fig. VI.1.1. To find the point p where the graph crosses the x-axis, execute:

```
Solve(f(p) = 0, p = 1)
```

The result is p = 0.876726. Now the integral can be computed. We use the command that produced Fig. VI.1.1:

```
Integ(f(x), x = 0 to p)
```

The result is 0.135698.

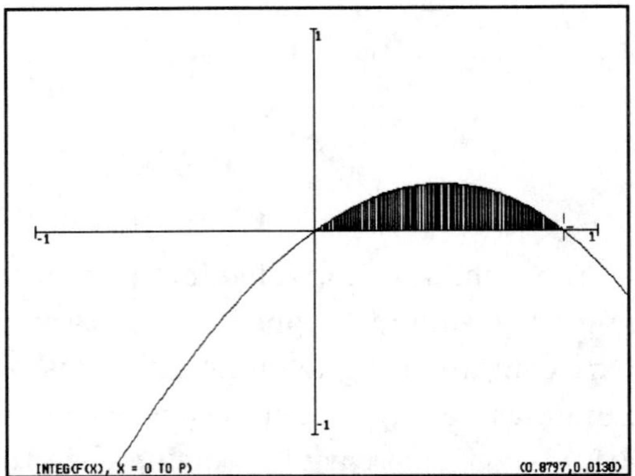

Fig. VI.1.1: Graph of y = sin(x) - x^2, highlighting area to be found.

VI.1.2 Finding a Maximum

Problem: Find the maximum of the function y = sin(x) - x^2.

Solution: We present three methods for finding the maximum: graphing, tabulating, and using the derivative. First define the function:

```
f(x) = sin(x) - x^2
```

Method 1: Locate the maximum of the function by graphing it and using the crosshairs:

```
Window(-1,1,-1,1)
Graph(f(x),x)
```

The graph is Fig. VI.1.2. The graphical estimate can be improved by examining a smaller window:

```
Window(0.44,0.47, 0.23,0.235)
Graph(f(x),x)
```

This graph appears in Fig. VI.1.3. The crosshairs show the maximum at (0.451, 0.233).

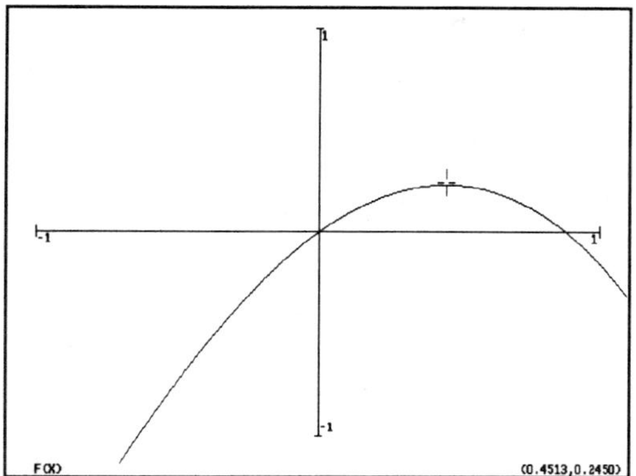

Fig. VI.1.2: Graph of $y = \sin(x) - x^2$, with crosshairs at maximum. Crosshair coordinates are (0.45, 0.24).

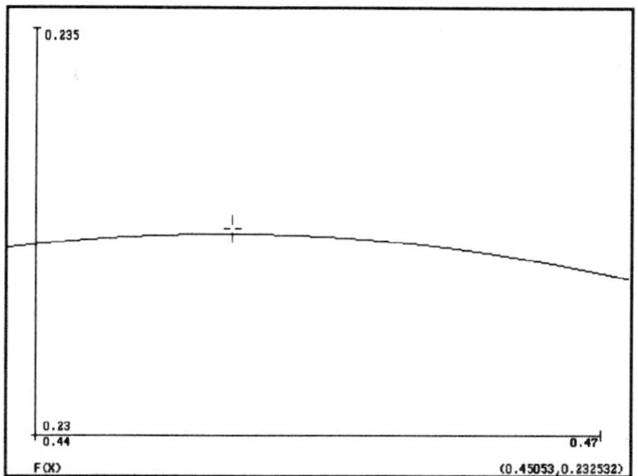

Fig. VI.1.3: Graph of $y = \sin(x)-x^2$ at higher resolution. Crosshairs show maximum at (0.451, 0.233).

Method 2: Use a table of function values:

```
table(f(x),x = 0 to 1 step 0.05)
```

CC displays the table:

x	f(x)	x	f(x)
0	0	0.55	0.220187
0.05	0.0474792	0.6	0.204642
0.1	0.0898334	0.65	0.182686
0.15	0.126938	0.7	0.154218
0.2	0.158669	0.75	0.119139
0.25	0.184904	0.8	0.0773561
0.3	0.20552	0.85	0.0287804
0.35	0.220398	0.9	-0.0266731
0.4	0.229418	0.95	-0.0890845
0.45	0.232466	1.0	-0.158529
0.5	0.229426		

Table of values for $f(x) = \sin(x) - x^2$

From this table, we see that the maximum occurs between $x = 0.4$ and $x = 0.5$. If you use the `Table` command to investigate this region:

```
table(f(x),x = 0.4 to 0.5 step 0.005)
```

you will see that the maximum of $f(x) = 0.232$ occurs near $x = 0.450$.

Method 3: Solve $f'(x) = 0$, like they taught you in calculus class:

```
Solve(dif(f(x),x)=0; x = 0.45)
y = f(x)
```

The result is that the maximum is located at $x = 0.450184$, and it is $y = 0.232466$.

VI.1.3 Finding Slopes and Graphing Tangents

Problem: Graph the tangent line to the graph of $y = \sin(x) - x^2$ where the curve crosses the x-axis to the right of the origin.

Solution: First we define and graph the function:

```
f(x) = sin(x) - x^2
Window(-1,1,-1,1)
QuickG(f(x),x)
```

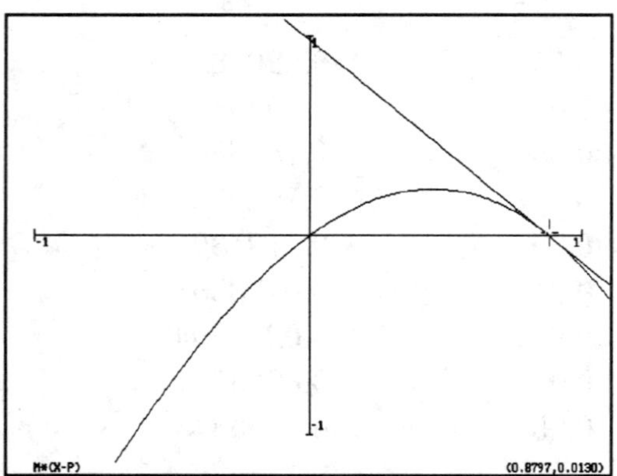

Fig. VI.1.4: Graph of $y = \sin(x)-x^2$ with tangent line at x-intercept.

Next we find the point p where the curve crosses the x-axis:

```
solve(f(p) = 0, p = 1)
```

The intersection is $p = 0.876726$, and the slope at p is $m = f'(p)$:

```
m = dif(f(x), x = p)
```

The slope is $m = -1.113782$, and the equation of the tangent line is $y = m(x-p)$. Add this line to the graph with the command:

```
sketch(m*(x-p),x)
```

The resulting figure is shown in Fig. VI.1.4

VI.1.4 Area between Curves

Problem: Find the area in the first quadrant between the curves $y = x$ and $y = 4\sin(x)$.

Solution: Begin by defining and graphing the functions $y = x$ and $y = 4\sin(x)$:

```
Keepline
Window(0,5,0,4)
QuickG(4*sin(x),x)
Sketch(x,x)
```

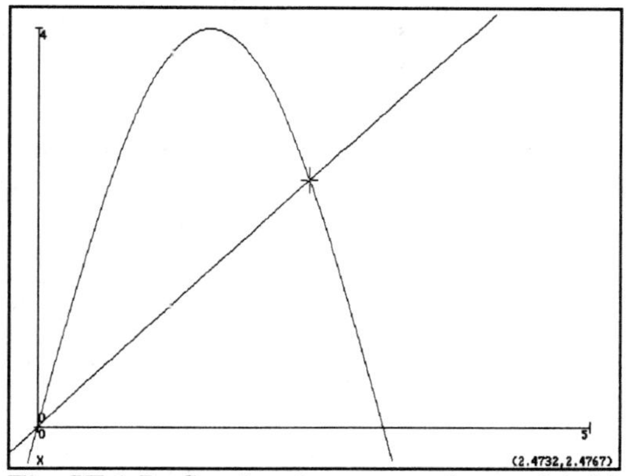

Fig. VI.1.5: Graphs of $y = x$ and $y = 4\sin(x)$

The right-hand limit of the area is the x-coordinate p where the curves cross. By placing the crosshairs where the curves intersect, we see that p ≈ 2.5. A more exact value for p may then be found:

```
Solve(p = 4*sin(p), p=2.5)
```

The exact result is $p = 2.474577$. Now we can compute the area as an integral:

```
In(4*Sin(x)-x, x = 0 to p)
```

The area is 4.08092.

VI.1.5 Study of an Ellipse

Problem: Graph the ellipse $3x^2 + 4y^2 = 12$, find the foci, and demonstrate that the sum of the distances from the foci to the ellipse is constant. Show that the area of the ellipse equals $\pi \cdot$*major axis minor axis*. Find the arc length of the ellipse.

Solution: We begin by graphing the ellipse two ways. First we create functions for the upper and lower halves of the ellipse and graph them, then we graph the ellipse as a parametric equation.

```
// Method 1:  graph +/- Sqrt((12 - 3x^2)/4)
Window(-2.8,2.8,-2,2)
Graph(Sqrt(3 - 0.75*x^2),x)
Graph(-Sqrt(3 - 0.75*x^2),x)

// Method 2: use parametric equations
x(t) = 2*Cos(t)             // parametric equations
y(t) = Sqrt(3)*Sin(t)       // for ellipse
Window(-2.8,2.8,-2,2)
Paramg(x(t),y(t),t = 0 to 2*pi)
```

(Fig. VI.1.6). To find the foci of the ellipse, we use the formula:

f = distance center to foci
R = major semiaxis
r = minor semiaxis

$$f^2 + r^2 = R^2$$

In our example, $r = \sqrt{3}$, $R = 2$, so we may enter the distance $f = 1$ from the center to the focus:

```
f = 1
```

Next we show that the sum of the distances from the foci to any point on the ellipse is 4. We begin by drawing lines from the foci to a point on the curve (Fig. VI.1.6):

```
line(f,0,x(pi/4),y(pi/4))
line(-f,0,x(pi/4),y(pi/4))
```

Using the parametric definition of the ellipse, we next define a function that gives the distance from the point ($x(t)$, $y(t)$) to the foci:

```
distance(t) =
   sqrt((x(t)-f)^2 + y(t)^2)
   + sqrt((x(t)+f)^2 + y(t)^2)
```

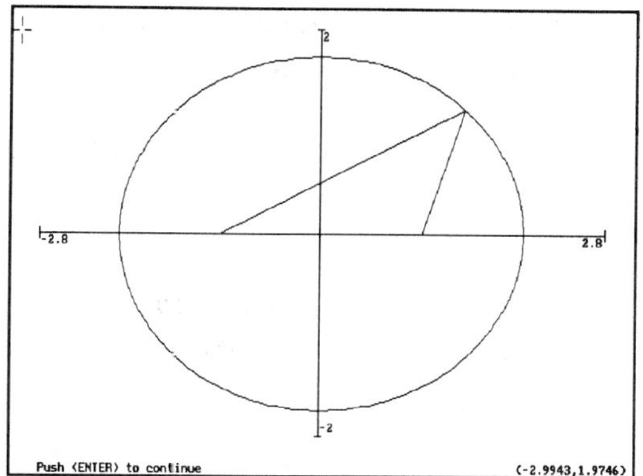

Fig. VI.1.6: Ellipse $3x^2 + 4y^2 = 12$, with lines drawn from the foci to a point on the curve. The sum of lengths of the lines is 4.

Now evaluate the distance function and graph it:

```
d = distance(1.5)
window(0,2*pi,0,5)
sketch(distance(t),t)
```

The graph of the distance function (Fig. VI.1.7) is constant, demonstrating that the distance from the foci to any point on the ellipse is 4.

The area of the ellipse is:

$$2\int_{-2}^{2}\frac{\sqrt{12-3x^2}}{2}\,dx$$

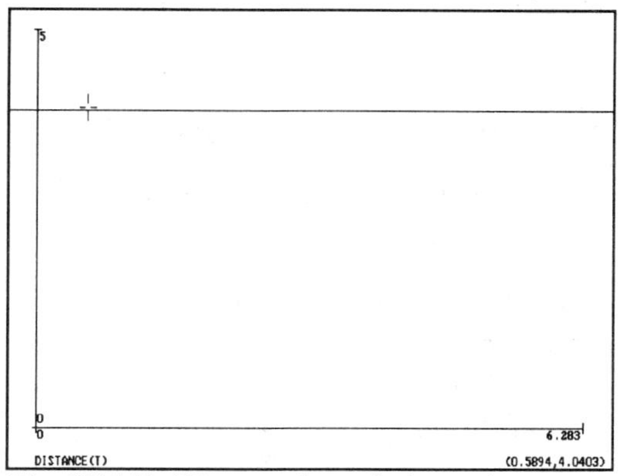

Fig. VI.1.7: Distance from foci to point ($x(t)$, $y(t)$), a constant.

which is computed:

```
area1 = 2*In(Sqrt(12 - 3*x^2)/2, x = -2 to 2)
```

The result is 10.88279. This is very close to π·*major axis minor axis*:

```
area2 = pi*2*sqrt(3)
```

which CC computes as 10.88280.

To find the arc length of the ellipse, we use the parametric representation of the curve and the formula:

$$arc\text{-}length = \int \sqrt{x'(t)^2 + y'(t)^2}\, dt$$

In CC

```
arclength = in(sqrt(dif(x(t),t)^2 + dif(y(t),t)^2),
  t = 0 to 2*pi)
```

The result is 11.7397.

VI.1.6 Computing Limits

Problem: Evaluate the renowned limit: $\lim_{x \to 0} \frac{\sin(x)}{x}$.

Solution: We begin by graphing the function:

```
Window(-2*pi,2*pi,-1,1)
Graph(sin(x)/x,x)
```

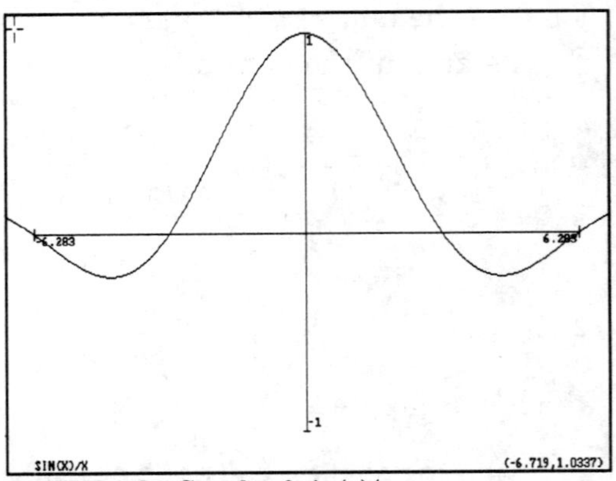

Fig. VI.1.8: Graph of sin(x)/x

(Fig. VI.1.8). Graphically, as x approaches 0, it appears that sin(x)/x approaches 1. We can check this by examining sin(x)/x for small values of x.

```
f(x)  = sin(x)/x
a1 = f(0.1)       // 0.998334
a2 = f(0.01)      // 0.999983
a3 = f(0.001)     // 1
a4 = f(-0.001     // 1
```

Again, the limit appears to be 1. Since sin(x) and x both equal 0 when x = 0, this limit can be computed with L'Hopital's rule:

```
Limit = dif(sin(x),x=0)/dif(x,x=0)    // limit is 1
```

(In all fairness, we should point out that L'Hopital's rule is perhaps inappropriate for this problem. L'Hopital's rule uses the derivative of sin(x), and the derivative of sin(x) in turn depends on evaluating the limit in question!)

See VI.1.11 for a further discussion of limits.

VI.1.7 The Cycloid

Problem: Find the arc-length and the area for one arch of the cycloid:

$$x = t - \sin(t)$$
$$y = 1 - \cos(t)$$

Solution: Enter the parametric equations for the cycloid into CC with the commands:

```
x(t) = t - sin(t)
y(t) = 1 - cos(t)
```

We can use these to graph the cycloid (Fig. VI.1.9):

```
Window(0,14,-1,3)
ParamG(x(t),y(t),t=-1 to
  15, 800)
```

To find the arc-length of this curve, use the integral:

$$\int_0^{2\pi} \sqrt{x'(t)^2 + y'(t)^2}\, dt$$

Enter this formula into CC as:

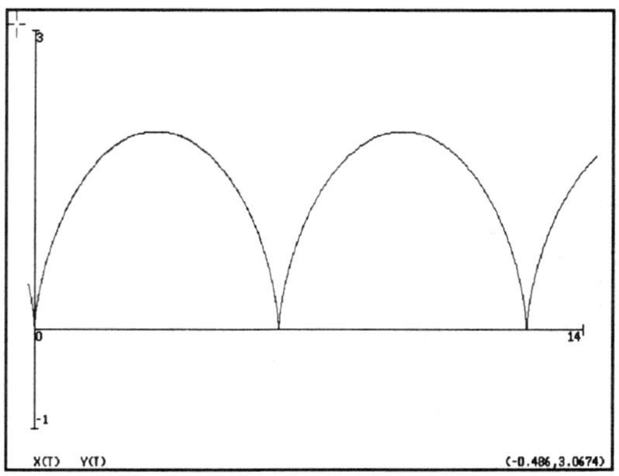

Fig. VI.1.9: Graph of the cycloid
$x = t - \sin(t)$, $y = 1 - \cos(t)$

```
Arc =in(sqrt(dif(x(t),t)^2 + dif(y(t),t)^2),t = 0 to 2*pi)
```

The result is *Arc* = 8.

To find the area under one arch of the cycloid, use the integral:

$$\int_{x=0}^{x=2\pi} y\,dx = \int_{t=0}^{t=2\pi} y(t)\,x'(t)\,dt$$

In CC, this becomes:

```
Area = In(y(t)*Dif(x(t),t),t = 0 to 2*pi)
```

The value is *Area* = 9.424778. You can check directly that this is equal to 3π.

VI.1.8 Graphing Hyperbolas and their Asymptotes

Problem: Graph the hyperbola $2x^2 - 3y^2 = 3$ and its asymptotes.

Solution: Begin by graphing the two branches of the hyperbola $y = \pm\frac{\sqrt{2x^2-3}}{3}$:

```
Window(-4.2, 4.2, -3, 3)
Linestyle(0)
KeepLine // use solid lines
Graph(Sqrt((2*x^2-3)/3),x)
Graph(-Sqrt((2*x^2-3)/3),x)
```

The asymptotes satisfy the equation: $2x^2 - 3y^2 = 0$ (compare to the equation of the hyperbola), or $y = \pm\sqrt{2/3}\ x$. We graph these lines with the commands:

```
Sketch(sqrt(2/3)*x,x)
Sketch(-sqrt(2/3)*x,x)
```

See Fig. VI.1.10 for the result.

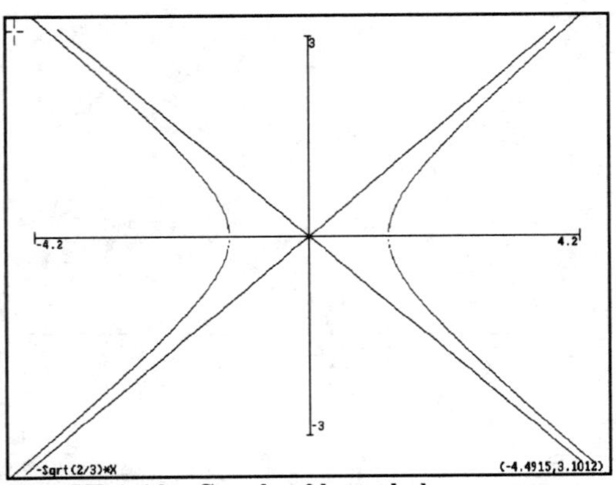

Fig. VI.1.10: Graph of hyperbola $2x^2 - 3y^2 = 3$ and its asymptotes.

VI.1.9 Areas Bounded by Parametric Curves

Problem: Find the area bounded by the parametric curve $y = t^3 - t$, $x = t^2 - 1$.

Solution: First, define the parametric equations and graph the curve (Fig. VI.1.11):

```
x(t) = t^2 - 1
y(t) = t^3 - t
Paramg(x(t),y(t),t=-2,2)
```

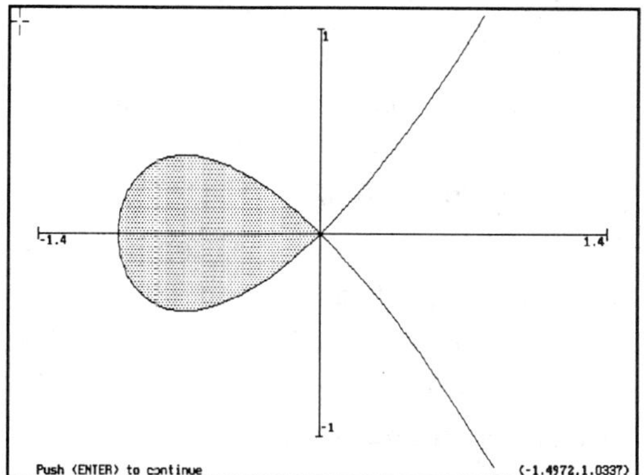

Fig. VI.1.11: Graph of $y = t^3 - t$, $x = t^2 - 1$ showing area to be found.

The upper curve bounding the area shown in the figure goes left to right from (x(0), y(0)) to (x(-1), y(-1)). To compute this area,use the equation from VI.1.7:

```
area = 2*In(y(t)*Dif(x(t),t),t=0 to -1)
```

The result is 0.533333.

VI.1.10 Inflection Points

Problem: Find the points of inflection on the curve $y^2 = x^3 - x$.

Solution: Begin by graphing one branch of the curve $y=\pm\sqrt{x^3 - x}$ and its second derivative (Fig. VI.1.12):

```
f(x) = sqrt(x^3-x)
window(0,3,-10,10)
quickg(f(x),x)
g(x) = dif(f(x),x,2)
quickg(g(x),x)
```

The inflection point occurs where $f''(x) = 0$. The graph of $f''(x)$ (also in Fig. VI.1.12) shows that the choice of first guess will be crucial for successfully finding its zero, and the cross-hairs at the root of $f''(x)$ show that the root is approximately 1.47. Using this initial approximation, CC will return a more accurate value for the x-coordinate of the inflection point:

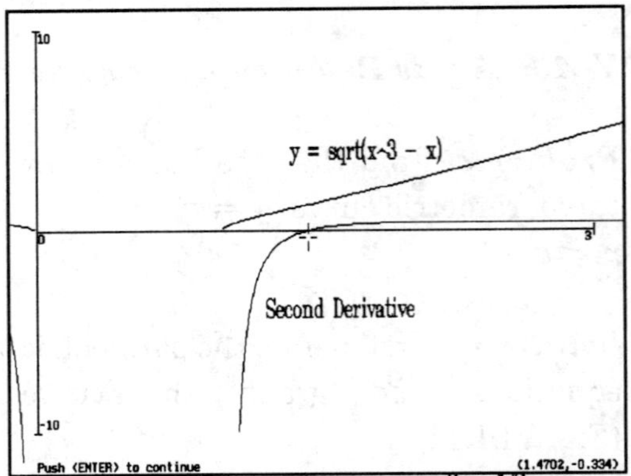

Fig. VI.1.12: The graph of $y=(x^3\text{-x})^{1/2}$ and its second derivative.

```
Solve(g(p) = 0, p=1.47) // p = x-coordinate of inflection point
f(p)                    // y-coordinate of inflection point
```

CC returns $p = 1.46789$ and $f(p) = 1.301911$.

VI.1.11 Improper Integrals

Problem: Estimate $\int_0^{\infty} \frac{1}{1+x^3} dx$ and find an error bound for your estimate.

Solution: *Always* begin by defining and graphing the function (Fig. VI.1.13):

```
f(x) = 1/(1+x^3)
window(0,10,0,1)
graph(f(x),x)
```

Since the accuracy of CC's integration algorithm decreases as the base length increases, we will split the integral into two parts:

$$g(a) = \int_0^a f(x)\,dx$$
$$h(a) = \int_a^\infty f(x)\,dx$$

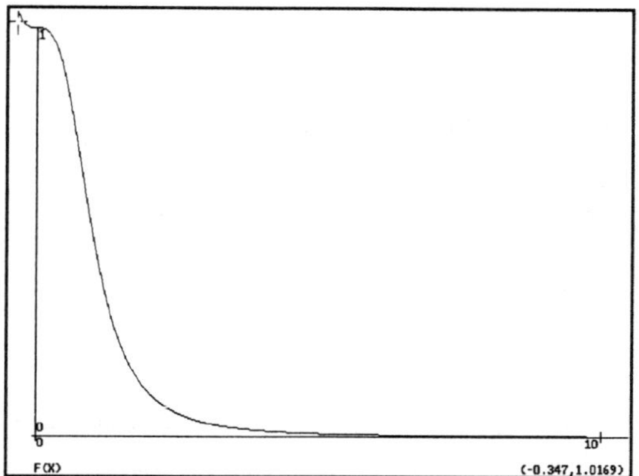

Fig. VI.1.13: Graph of $y = 1/(1 + x^3)$.

We can define $g(a)$ directly in CC, but estimating $h(a)$ will take more thought.

```
g(a)  =  In(f(x),x=0 to a)
```

The integral $g(a)$ is an elliptic integral which cannot be evaluated exactly but which can be evaluated numerically to high accuracy for reasonable values of a. Evaluating $h(a)$, on the other hand, is best done by finding integrals which can be evaluated exactly and which are upper and lower bounds for $h(a)$. We note that:

$$\begin{aligned}\frac{1}{x^3} &> f(x)\\ &= \frac{1}{x^3} - \left(\frac{1}{x^3} - f(x)\right)\\ &= \frac{1}{x^3} - \frac{1}{x^3 + x^6}\\ &> \frac{1}{x^3} - \frac{1}{x^6}\end{aligned}$$

Since $\int_a^\infty \frac{1}{x^n}\,dx = \frac{1}{(n-1)\,a^{n-1}}$, we may integrate this inequality to obtain bounds for $h(a)$:

$$\frac{1}{2a^2} - \frac{1}{5a^5} < h(a) < \frac{1}{2a^2}$$

Now we can define an approximation for $h(a)$ midway between the bounds:

```
h1(a)  =  1/(2*a^2)-1/(10*a^5)
```

The entire improper integral is then approximated in CC with the equation:

```
q(a) = g(a)+h1(a)
```

with an error of less than $\frac{1}{10a^5}$ plus the error in computing the first integral $g(a)$. Some sample values are:

```
q1 = q(10)       // 1.209201
q2 = q(20)       // 1.209200
q3 = q(50)       // 1.209204
```

To obtain the best possible estimate of our integral, we should select a value for a minimizing the combined errors $1/(10a^5)$ + error in $g(x)$. That is, in principle we have $\lim_{a\to\infty} q(a) = \int_0^\infty \frac{1}{1+x^3}\,dx$. However, for large values of a, $g(a)$ will have so much error that $q(a)$ will not be the best value for the integral. A rough-and-ready procedure for computing the best approximation to the integral is to construct a sequence of values $a_i = q(5i)$, $i = 1,2,3, \ldots$ and, from a_i to construct the sequence of absolute differences $b_i = |a_{i+1} - a_i|$. As i increases, a_i will approach the limit and the absolute difference b_i will approach 0. However, when i is so large that the error in the integral algorithm starts to grow, the sequence b_i will start to increase. The best approximation to the integral is a_{i+1}, where i is chosen to minimize b_i.

We carry out this procedure in CC:

```
a = list(q(n),n=5 to 50 step 5);
b = list(abs(a[i+1]-a[i]),i=1,9)
m = min(b)                // minimum value is b[7]
```

The values of a_i and b_i are displayed below. Since the minimum entry in the sequence of b's is b_7 = 0.00000004451794666, the best approximation for the integral is a_8 = 1.209199668. Because CC only uses about 12 digits accuracy, and because b_7 is the difference of two much larger values $a_7 - a_8$, only the first few digits of b_7 are meaningful. Nevertheless, the magnitude of b_7 indicates that the integral in question equals 1.209200 up to seven digits accuracy.

i	a_i	b_i
1	1.2092312674	0.00003047524115
2	1.2092007921	0.0009510843647
3	1.2101518765	0.0009517302897
4	1.2091998869	0.0000002593242243
5	1.2092001462	0.00000008511960914
6	1.2091998018	0.0000000892287062
7	1.2091997125	0.00000004451794666
8	1.209199668	0.000004613242709
9	1.2092042813	0.0000009937539289
10	1.2092032875	

VI.1.12 Graphing Tabular Data

Problem: Given a collection of paired values (a_i, b_i), make a line graph connecting the data points.

Solution: The data we use comes from a report prepared by the author's seventh-grade son. The data represent years and the number of automobiles sold by the leading manufacturer in that year. These data are entered into a pair of lists in CC:

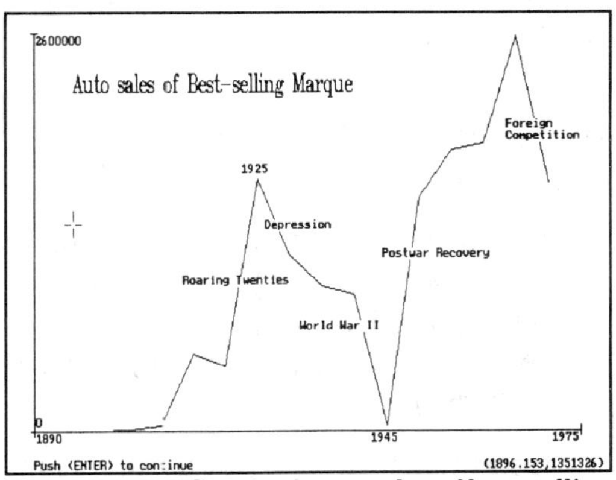

Fig. VI.1.14: Graph of auto sales of best-selling marques every five years.

```
// Enter the dates
years = List(j,j=1900 to 1970 step 5)

// Enter the sales for each year
sales = (1500,6500,32053,501462,419517,1643295,1155162,942439,
895734,35439,1520577,1830029,1873598,2587490,1621846)

// make the graph
Window(1890,1975,0,2600000)
ListG(years, sales)
```

The labels were added with the `Write@C` command.

VI.1.13 Self-Intersection of Lissajous Curve

Problem: Find the exact coordinates of the self-intersection of the Lissajous curve $x = \sin(6t), y = \sin(7t)$ nearest (-0.63, 0.70). Determine the angle of self-intersection at this point.

Solution: As always, begin by graphing the curve in question (Fig. VI.1.15):

```
Window(-1.4,1.4,-1,1)
x(t) = sin(6*t)
y(t) = sin(7*t)
ParamG(sin(x(t),y(t),t=0 to
  2*pi)
```

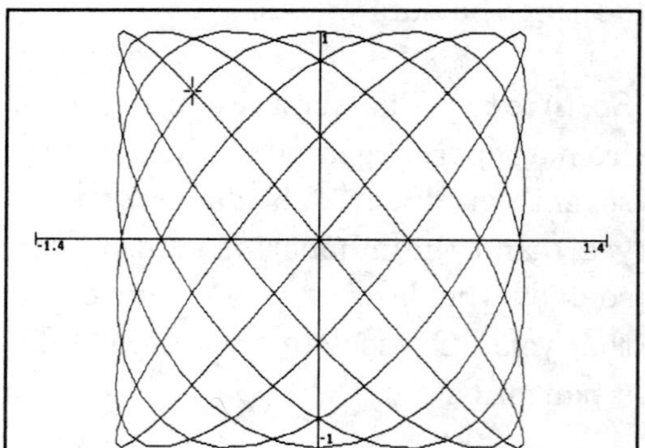

Fig. VI.1.15: Lissajous curve x = Sin($6t$), y = Sin($7t$) with crosshair at self-intersection to be found.

To find the coordinates of the point in question, we will find the two values of the parameter t in the range $[0, 2\pi]$ that yield this point. If we call these values t_1 and t_2, then t_1 and t_2 satisfy the equations:

$$x(t_1) = x(t_2)$$
$$y(t_1) = y(t_2)$$

Any solution to this system with $t_1 \neq t_2$ corresponds to a self-intersection of the Lissajous curve. We wish to find a self-intersection near (-0.63, 0.70). That is, our points t_1 and t_2 must both be approximate solutions to the system of equations:

$$x(t) + 0.63 = 0$$
$$y(t) - 0.70 = 0$$

After graphing both these equations (Fig. VI.1.16):

```
window(0,2*pi,-2,2)
graph(x(t)+0.63,t)
graph(y(t)-0.70,t)
```

the crosshairs show that the common solutions occur at $t_1 = 3.02$, $t_2 = 4.83$. Using these values as starting approximations, we can solve the first pair of equations for the two parameter values t_1 and t_2 corresponding to the desired self-intersection of the Lissajous curve:

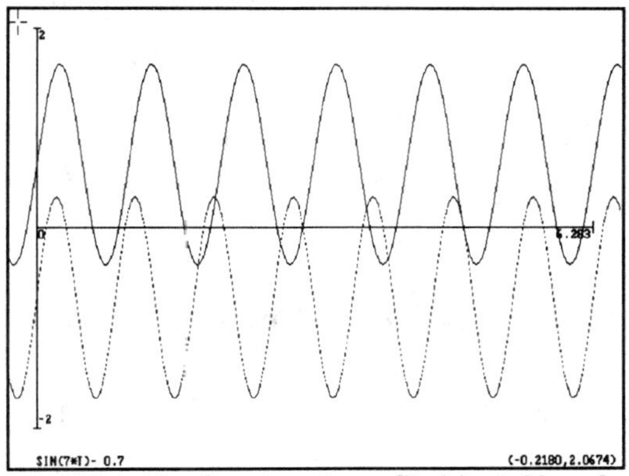

Fig. VI.1.16: Graphs of y = Sin(x)+0.63 and y = Sin(x)-0.70.

```
Solve(x(t1)=x(t2), y(t1)=y(t2), t1 = 3.02, t2=4.83)
a=x(t1)
b=y(t1)
```

The result is $t_1 = 3.029393$, $t_2 = 4.824589$, $a = -0.62349$, $b = 0.707107$. The intersection point is (a, b). To check this result, compare $(x(t_1), y(t_1))$ to $(x(t_2), y(t_2))$:

```
a-x(t2)    // should be near 0
b-y(t2)    // also near 0
```

To find the angle at which the Lissajous curve intersects itself, we first define some general purpose vector functions: the dot product, the Euclidean norm and the angle between two vectors:

```
dot(v,w)   = Sum(v*w)
norm(v)   = Sqrt(dot(v,v))
angle(v,w)   = ACos(dot(v/norm(v),w/norm(w)))
```

Next we express the Lissajous curve as the range of a vector-valued function **f**(t):

```
f(t)  =  (x(t), y(t))
```

Since **f**(t) is a vector-valued function, its derivative is a vector too, the tangent vector to the Lissajous curve. The angle at the point of self-intersection is the angle between the tangent vectors **f'**(t_1) and **f'**(t_2):

c = angle(dif(f(t),t=t1),dif(f(t),t=t2))

The result is 1.5617129 radians. Notice that CC computed the vector derivatives of the vector-valued function **f**(t) at $t = t_1$ and $t = t_2$.

VI.1.14 Midpoint, Trapezoidal and Simpson's Approximations to Integrals

Problem: Calculate and compare the midpoint, trapezoidal and Simpson's approximations for the integral $\int_{-1}^{2} \frac{1}{1+x^4} dx$.

Solution: Begin by defining the function to integrate:

```
f(x)  =  1/(1+x^4)
```

The n'th midpoint approximation to $\int_a^b f(x)\, dx$ is:

$$dx = \frac{b-a}{n}$$

$$x_i = a + \left(i + \frac{1}{2}\right) dx \qquad i = 0 \ldots n$$

$$midpointapproximation = \left(\sum_{i=0}^{n-1} f(x_i) \right) dx$$

In CC, the n'th midpoint approximation to the integral of $f(x)$ on the interval (a, b) is:

```
dx = (b-a)/n
midpt(a,b,n) = sum(f(x),x=a+dx/2 to b-dx/2 step dx)*dx
```

The n'th trapezoidal approximation to $\int_a^b f(x)\,dx$ is:

$$dx = \frac{b-a}{n}$$
$$x_i = a + i\,dx \qquad i = 1 \ldots n-1$$
$$\textit{Trapezoidal Approximation} = \left(\frac{f(a)}{2} + \sum_{i=1}^{n-1} f(x_i) + \frac{f(b)}{2}\right)dx$$

In CC, the n'th trapezoidal approximation to the integral of $f(x)$ on the interval (a,b) is:

```
dx = (b-a)/n
Trap(a,b,n) = (f(a)/2 + sum(f(x),x=a+dx to b-dx step dx)
  + f(b)/2)*dx
```

To examine the midpoint and trapezoidal approximations for n = 4, 6, ..., 30 to our integral, we use the `List` constructor to collect all these values:

```
m = list(midpt(-1,2,n),n=4 to 30 step 2)
t = list(trap(-1,2,n),n=4 to 30 step 2)
```

These values are presented in the table below.

Finally, the n'th Simpson's approximation to the integral is most easily defined as:

Simpson's Approximation
= (1/3)*Trapezoidal Approximation* + (2/3)*Midpoint Approximation*

We can use the lists of values already created to compute a list of values for Simpson's approximation:

```
s = (2m+t)/3
```

The values produced are:

n	Midpoint	Trapezoidal	Simpson's
4	1.967429	1.880457	1.938438
6	1.949217	1.913357	1.937263
8	1.943732	1.923943	1.937136
10	1.941318	1.928711	1.937115
12	1.940018	1.931287	1.937108
14	1.939239	1.932835	1.937105
16	1.938736	1.933837	1.937103
18	1.938391	1.934524	1.937102
20	1.938145	1.935014	1.937102
22	1.937964	1.935377	1.937101
24	1.937825	1.935653	1.937101
26	1.937718	1.935867	1.937101
28	1.937633	1.936037	1.937101
30	1.937564	1.936174	1.937101

You can see from the table how the midpoint and trapezoidal approximations converge slowly towards a common value, and much more rapidly Simpson's approximations converge. These values are graphed in Fig. VI.1.17, which was constructed with the commands:

```
Window(3,31,1.87,1.97)
j := List(i, i=4 to 30 step 2)
ListG(j,m)
ListG(j,t)
ListG(j,s)
```

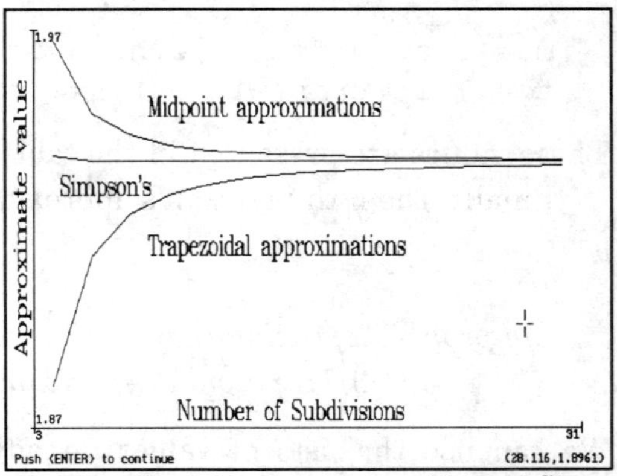

Fig. VI.1.17: Midpoint, Trapezoidal and Simpson's Approximations to an integral.

VI.1.15 Self-Intersecting Polygons

Problem: Graph self-intersecting regular polygons.

Solution: We will graph polygons with n sides, with each side going from one vertex to the m'th following vertex. We require that m and n be relatively prime. The CC command that draws this polygon is:

```
PolarG(1, θ=0 to 2mπ step n)
```

Of course you can't use this command without first specifying m and n. The figures in this section show the effects of using different values for m and n.

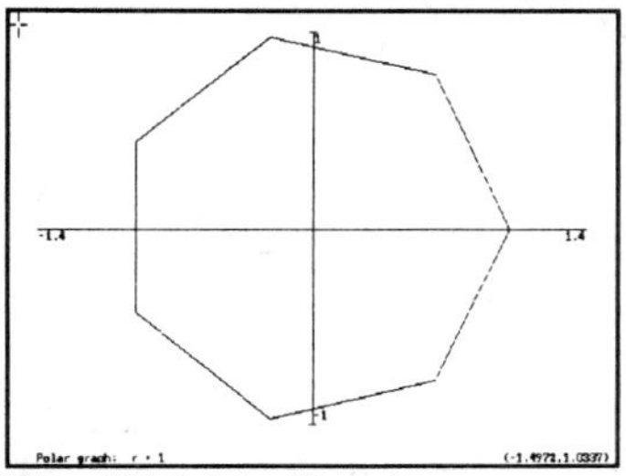

Fig. VI.1.18: Regular 7-gon: $m = 1, n = 7$.

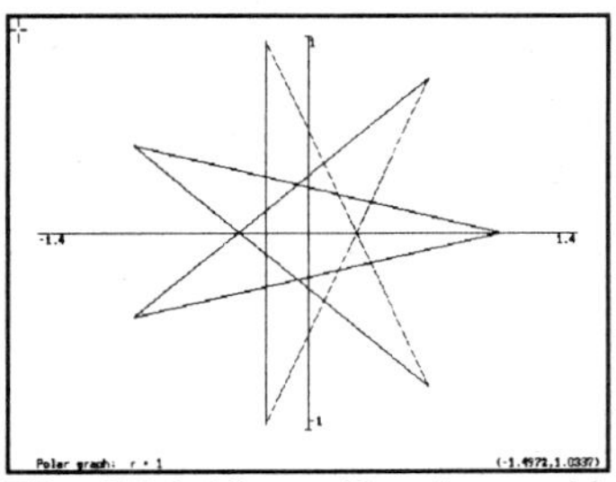

Fig. VI.1.19: Regular self-intersecting 7-gon: m=3, n=7.

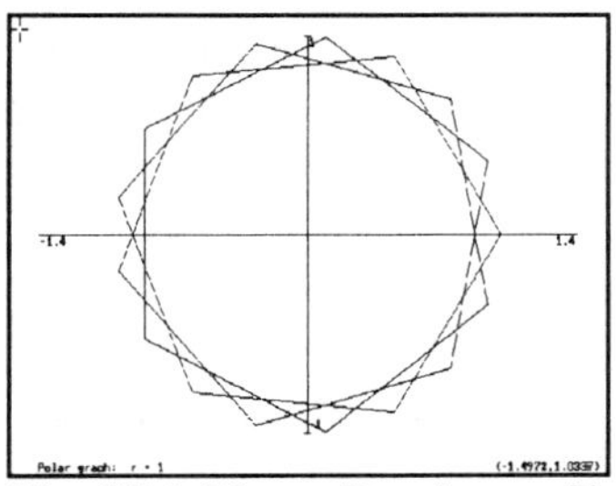

Fig. VI.1.20: Regular self-intersecting 17-gon: m=3, n=17.

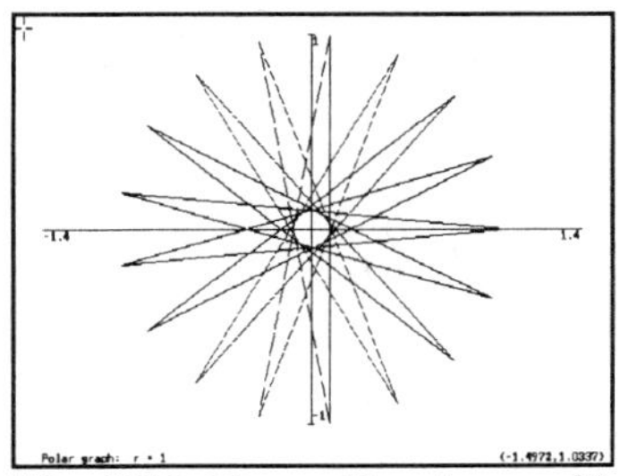

Fig. VI.1.21: Regular self-intersecting 17-gon: m=8, n=17.

VI.1.16 Taylor Series

Problem: Compute and graph the n'th Taylor series approximations about 0 to Cos(x) for n = 3, 9, 15 and 21.

Solution: The Taylor polynomial about 0 of degree n for an arbitrary function $f(x)$ is:

$$n'th\ degree\ Taylor\ polynomial = \sum_{i=0}^{n} \frac{f^{(i)}(0)x^i}{i!}$$

In CC the Taylor polynomial is defined:

```
Taylor(x,n)  = Sum(Dif(f(t),t=0,i)*x^i/i!,i=0 to n)
```

To apply this operator to Cos(x) and graph the result, enter the formulas:

```
f(x)  = cos(x)                 // Define the function to examine
Window(0,3pi,-1,1)             // Define graphing window
Graph(f(x),x)                  // Graph Cos(x) to compare to Taylor
Taylor(x,5)                    // View symbolic Taylor polynomial
Linestyle(0)                   // Make all graphs solid lines
Keepline
Graph(Taylor(x,3),x)           // Graph 3'rd Taylor approximation
Graph(Taylor(x,9),x)           // Graph 9'th Taylor approximation
QuickG(Taylor(x,15),x)         // Quick Graph 16'th Taylor approx
QuickG(Taylor(x,21),x)         // Quick Graph 21'st Taylor approx
```

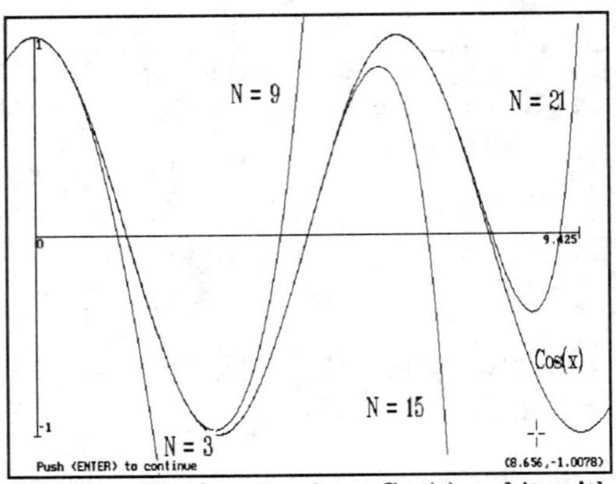

Fig. VI.1.22: Graphs of y = Cos(x) and its n'th Taylor approximations.

VI.1.17 Parallel Curves

This problem was developed by J. T. Smith of San Francisco State University. He calls it "the onset of kinkiness."

Problem: Construct a curve at a constant distance from the parabola $y = x^2$. Which of these curves are smooth?

Solution: We begin with a general solution to the problem. Given a parametric curve **C**(t) = $(x(t), y(t))$ without singularities, we will construct the parallel curve at distance w and determine when the parallel curve is singular.

The components of the tangent vector to **C**(t) are:

```
dx(t)  = dif(x(t),t)
dy(t)  = dif(y(t),t)
```

The unit normal vector **N**(t) to **C**(t) is then:

```
N(t)  = (-dy(t),dx(t))/sqrt(dy(t)²+dx(t)²)
```

Note that **N**(t) is a vector-valued function.

The vector equation **CP**(t) for the parallel curve is then:

```
CP(t)  = (x(t),y(t)) + w*N(t)
```

The curvature $k(t)$ for the original curve **C**(t) is:

```
ddx(t)  = dif(dx(t),t)
ddy(t)  = dif(dy(t),t)
k(t)  = (dx(t)ddy(t)+ddx(t)dy(t))/(dx(t)²+dy(t)²)^(3/2)
```

The reader can check the following calculation:

$$|\mathbf{CP}'(t)| = |\mathbf{C}'(t)|\,|1 - w\,k(t)|$$

Since the original curve is assumed non-singular, its derivative **C**'(t) never vanishes. Thus the derivative of the parallel **CP**(t) curve vanishes when $k(t) = 1/w$. For the parallel curve to be non-singular, we must choose w so that this equation fails for all t.

Let's do an example. The parabola is defined by the parametric equations:

```
x(t) = t
y(t) = t^2
```

First we graph the inverse curvature $1/k(t)$ (Fig. VI.1.23) to find the critical values of w:

```
Window(-1.4,1.4,0,3)
Graph(1/k(t),t)
```

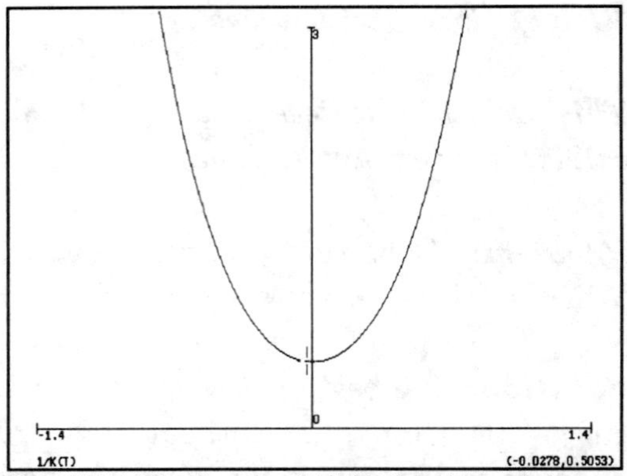

Fig. VI.1.23: Graph of $1/k(t)$ showing minimum of 0.5 at $t = 0$.

The minimum of $1/k(t)$ occurs at $t = 0$, where its value is:

```
m = 1/k(0)
```

We find $m = 0.5$, so if $w < 0.5$, the parallel curve will be non-singular. Let's graph some of them (Fig. VI.1.24):

```
Window(-1.25*1.4,1.25*1.4,
  -0.5,2)
Paramg(x(t),y(t),t=-1.4 to
  1.4)
w = -0.25
Paramg(CP(t)[1],CP(t)[2],
  t=-1.4 to 1.4)
```

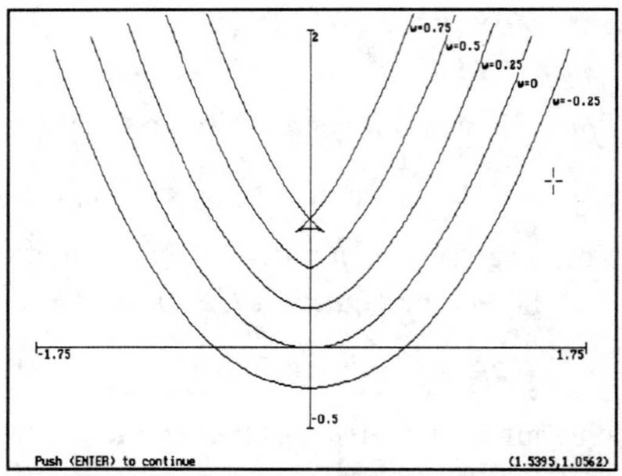

Fig. VI.1.24: Curves at distance w from $y = x^2$. When $w = 0$, we have the original curve; the curve is singular for $w \geq 0.5$

Redefine w to be 0.25, 0.5 and 0.75, and re-execute the last line to draw all the parallel curves. It is important to use a window with the correct aspect ratio, or the curves won't appear parallel on the screen.

VI.1.18 Cauchy Integrals

Problem: Evaluate $\oint \frac{1}{z}dz$ on a closed curve about 0 in the complex plane.

Solution: First parametrize the unit circle in the complex plane and integrate $1/z$ on the circle:

```
z(t) = exp(î*t)
In(1/z(t)*dif(z(t),t), t = 0 to 2*pi)
```

The result is exactly $2\pi i$, within CC's limits of accuracy.

For a second solution, integrate the complex variable z on a diamond around the origin:

```
In(1/z,z=1,î)+In(1/z,z=î,-1)+In(1/z,z=-1,-î)+In(1/z,z=-î,1)
```

The result is `-4.297362466E-11 + 6.28318530 59î`, or very nearly $2\pi i$. This example illustrates how CC integrates complex-valued functions.

VI.1.19 Tangent Lines to Polar Curves

Problem: Sketch the tangent line to the polar curve $r = 0.5 + \cos(\Theta)$ at $\Theta = \pi/2$.

Solution: First graph the polar curve:

```
f(Θ) = 1/2 + cos(Θ)
window(-1,3,-1.5,1.5)
PolarG(f(Θ),Θ=0 to 2pi)
```

The angle α between the tangent line to $r = f(\Theta)$ and the radial line at angle Θ is:

$$\alpha = \text{Tan}^{-1}(f(\Theta)/f'(\Theta))$$

In CC, taking $\Theta = \pi/4$:

```
α =
  atan(f(π/4)/dif(f(Θ),Θ=π/4))
```

The slope of the tangent line is then:

```
slope = tan(π/4 + α)
```

The (x,y)-coordinates of the point where the line will be tangent to the curve are:

```
x0 = f(π/4)*cos(π/4)
y0 = f(π/4)*sin(π/4)
```

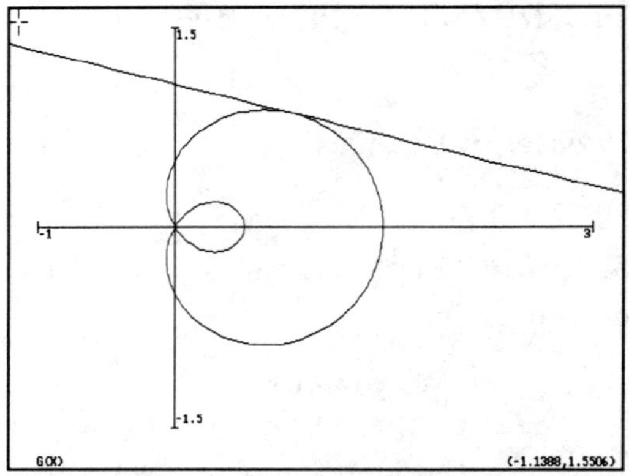

Fig. VI.1.25: Graph of $r = 0.5 + \mathrm{Cos}(\Theta)$ and tangent line at $\Theta = \pi/4$.

The equation of the tangent line is then:

```
g(x) = y0 + slope*(x-x0)
```

Add the tangent line to the graph (Fig. VI.1.25):

```
Sketch(g(x),x)
```

VI.1.20 The Hyperboloid of One Sheet as a Ruled Surface

This example was first worked out by Ralph Lakness of San Francisco State University.

Problem: Graph the hyperboloid of one sheet as a ruled surface.

Solution: Imagine a ring of radius 1 about the z-axis in $\mathbb{R}^3$ at height 1, and another at height -1. Connect these rings by vertical lines; the resulting figure is a cylinder. Twist the top ring and the cylinder become a hyperboloid of one sheet.

To draw a picture of this construction, begin by defining the two rings and the lines between them:

```
p1(Θ) = (cos(Θ), sin(Θ), -1)
p2(a,Θ) = (cos(a+Θ), sin(a+Θ), 1)
f(a,Θ,t) = p1(Θ) + t*(p2(a,Θ) - p1(Θ))
```

The function $p1(\Theta)$ parametrizes the bottom ring, $p2(a, \Theta)$ parametrizes the top ring twisted by an angle a, and $f(a, \Theta, t)$ parametrizes the line between $p1(\Theta)$ and $p2(a, \Theta)$. As t goes from 0 to 1, $f(a, \Theta, t)$ traverses a line from the bottom ring to the top; as Θ goes from 0 to 2π these lines sweep out the hyperboloid. To construct one of these figures, for example the one with $a = \pi/3$, use the command:

```
a = π/3
x(Θ,t) = f(a,Θ,t)[1]
y(Θ,t) = f(a,Θ,t)[2]
z(Θ,t) = f(a,Θ,t)[3]
ParamG3D(x(Θ,t),y(Θ,t),z(Θ,t); Θ=0 to 2π;
  t=0 to 1; -1,1; -1,1; -1,1; 20)
```

We have graphed the resulting figure for $a = 0$, $\pi/3$ and $2\pi/3$.

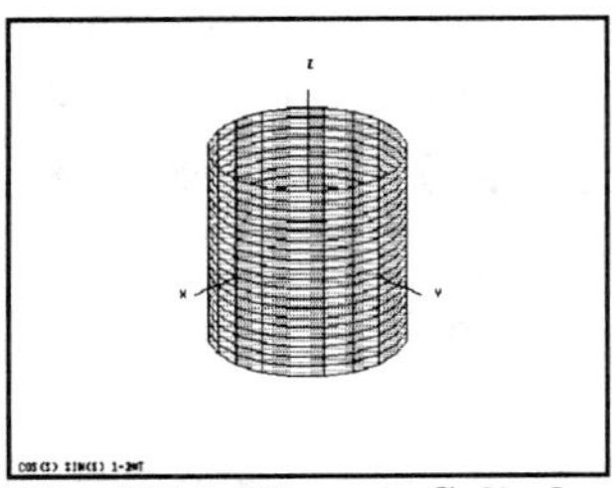

Fig. VI.1.26: Cylinder twisted by 0 radians.

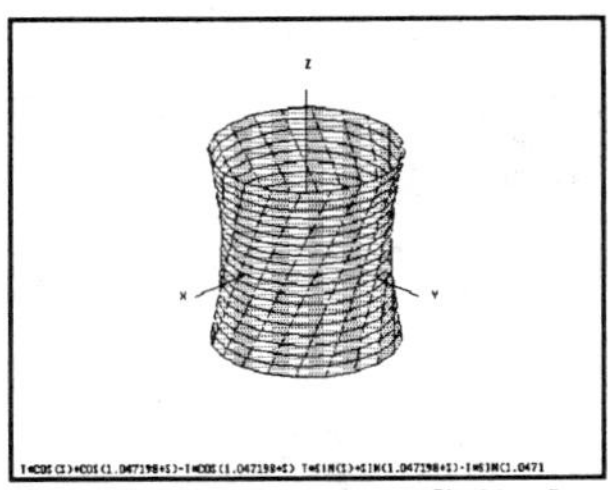

Fig. VI.1.27: Cylinder twisted by $\pi/3$ radians.

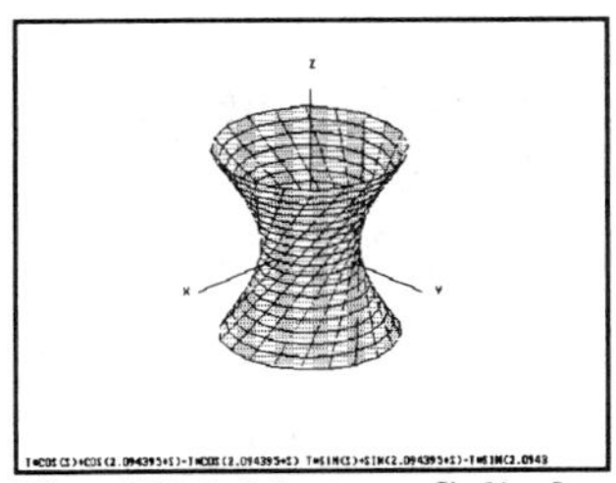

Fig. VI.1.28: Cylinder twisted by $2\pi/3$ radians.

Chapter 6
Applications

VI.1.21 Computing and Graphing Fourier Series

Problem: Find and graph the Fourier series for $y = \sin(x^2)$ on $[-\pi, \pi]$.

Solution: First define and graph the function in question:

```
f(x) = sin(x^2)
Window(-pi,pi,-1,1)
Graph(f(x),x)
```

The Fourier series for $f(x)$ on $[-\pi,\pi]$ is $\sum_i a_i \cos(ix)$, where:

$$a_0 = \frac{1}{2\pi}\int_{-\pi}^{\pi} f(x)\,dx$$

$$a_i = \frac{1}{\pi}\int_{-\pi}^{\pi} f(x)\cos(ix)\,dx \qquad i > 0$$

You can use CC to define a list of these Fourier coefficients. To speed the process up, use `InDepth` (III.12.5) to increase the speed (but reduce the accuracy) of CC's integrals. Since the factor *IntegrationTolerance* was not changed, CC's integration algorithm will not converge completely. The message "Possible inaccuracies ..." will be displayed, but the values returned will be more than sufficiently accurate for graphing purposes. `a0` is a_0, `a[i]` is a_i:

```
indepth(3)
a0 = 1/(2pi)*in(f(x),x=-pi,pi)
a = 1/pi*(list(in(cos(i*x)*f(x),x=-pi,pi),i=1,10))
```

Define the n'th partial Fourier series approximating $f(x)$:

```
g(x,n)=a0+sum(a[i]*cos(i*x),i=1,n)
```

Graphs for the 1'st, 4'th, 7'th and 10'th approximation appear in Fig. VI.1.29-Fig. VI.1.32.

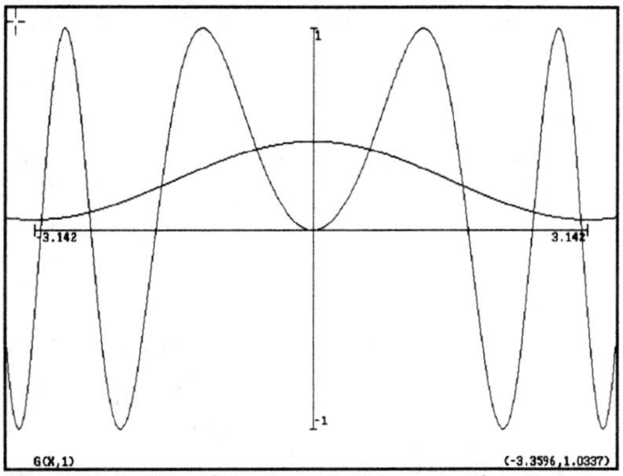

Fig. VI.1.29: Graphs of $f(x)$ and 1'st Fourier approximation.

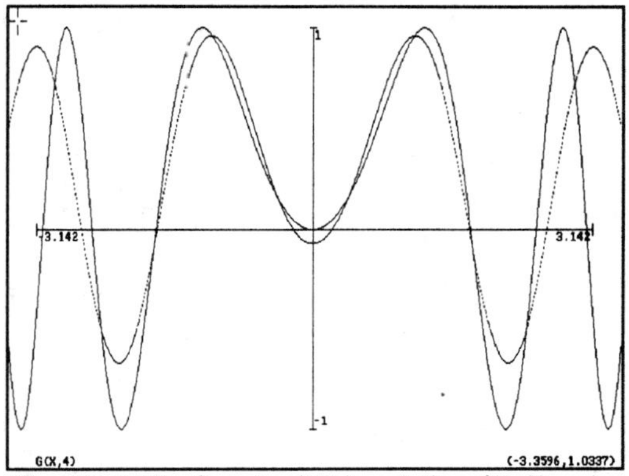

Fig. VI.1.30: Graph of $f(x)$ and the 4'th Fourier approximation.

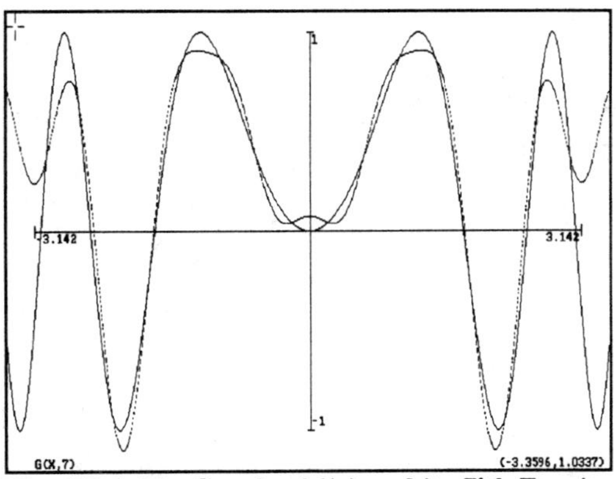

Fig. VI.1.31: Graph of f(x) and its 7'th Fourier approximation.

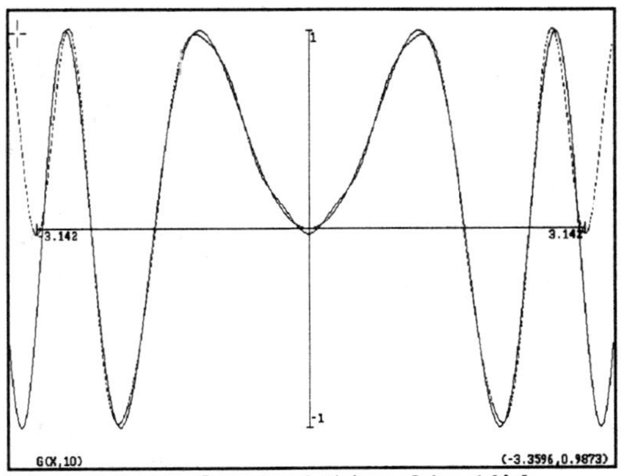

Fig. VI.1.32: Graph of $f(x)$ and its 10'th Fourier approximation.

VI.1.22 Financial Modeling

The following material demonstrates another potential use of CC. It is not intended as a financial recommendation. Not only have we failed to take any tax considerations into account, but the conclusions of the model are extremely sensitive to the assumptions. If the assumptions fail to hold in the future, so will the conclusion.

Chapter 6
Applications

Before making any financial decisions, you should consult a knowledgeable financial advisor.

***Problem*:** Determine if it is financially preferable to purchase a house or to rent a home and save the difference.

***Solution*:** We will compare the profit to be made through purchasing a home with a 30-year mortgage vs. renting a home and saving the difference between the monthly mortgage payment and the monthly rent. We begin by entering some assumptions: the house will cost $225,000, the down payment will be $40,000, the term of the mortgage will be 30 years, and the rent for the equivalent home will be $950/month:

```
HouseCost = 225000
DownPayment = 40000
Mortgage = HouseCost - DownPayment
Term = 360 // months
StartRent = 950
```

We must make some assumptions about the future; annual inflation rate will be 3%, the annual rate of return on savings will be 8.4%, and the annual mortgage rate will be 11%:

```
Inflation = 0.03
SaveRate = 0.084
MortgageRate = 0.11
```

Next we define some factors representing monthly growth factors for inflation, savings and mortgage debt:

```
I = 1 + Inflation/12
S = 1 + SaveRate/12
M = 1 + MortgageRate/12
```

If rent increases with inflation, then the rent for the *n*th month will be:

```
Rent(n) = StartRent*I^n
```

On the other hand, the monthly mortgage payment is a fixed amount given by:

```
MortgagePay = Mortgage*(M-1)/(1-M^(-Term))
```

(This formula is a standard one from annuity theory; see any text on business math.)

If you rent instead of purchase a home, in the *n*'th month you will be able to save:

```
Save(n) = MortgagePay - Rent(n)
```

The value of the house also increases with inflation, so that its value in the *n*'th month will be:

```
HouseValue(n) = HouseCost*I^n
```

The amount owed on your mortgage after the *n*'th month, and your equity in your home in the *n*'th month will be (according to standard annuity formulas):

```
Owed(n)=Mortgage*M^n-MortgagePay*(M^n-1)/(M-1)
Equity(n) = HouseValue(n) - Owed(n)
```

The value of your home, when you pay it off, will be:

```
FinalEquity = Equity(Term)
```

To compare the value of renting to the value of owning, we must compute

```
FVSave(n) = Save(n)*S^(Term-n)
TotalSave = DownPayment*S^Term + sum(FVSave(j),j=1,Term)
```

`FVSave(n)` is the future value, after 30 years, of the amount saved by renting instead of buying, in the *n*'th month; and `TotalSave` is the total savings, with accumulated interest, after 30 years.

Here are some sample results from these calculations:

Variable	*Value*	*Value*
HouseCost	225000	225000
DownPayment	40000	40000
Mortgage	185000	185000
Term	360	360
StartRent	950	950
Inflation	0.03	0.05
SaveRate	0.084	0.8
MortgageRate	0.11	0.12
MortgagePay	1763	1903
FinalEquity	552789	1005242
TotalSave	1254449	805408

Sample Values for Computation of House Rental vs. Purchase

As you can see, one scenario predicts that it is preferable to rent, the other that it is preferable to purchase a home.

VI.1.23 Discontinuous Functions

Problem: Construct the discontinuous functions:

$$Step(a, x) = \begin{cases} 1 & \text{if } x < a \\ -1 & \text{if } x > a \end{cases}$$

$$Char(a, b, x) = \begin{cases} 1 & \text{if } a < x < b \\ 0 & \text{if } x < a \text{ or } x > b \end{cases}$$

Solution: The functions are:

```
Step(a,x) = (x-a)/abs(x-a)
Char(a,b,x) = (Step(a,x) - Step(x,b))/2
```

These functions can be used to construct other discontinuous functions. For example, the square wave which is 1 in the domain $(2i, 2i+1)$ and -1 in the domain $(2i\text{-}1, 2i)$ can be constructed on the (large) interval (n, m), assuming n and m are integers:

```
Square(x) = sum(Par(i)*Char(i,i+1,x),i = n to m)
```

VI.2 Programming Applications

The applications which follow illustrate the uses of CC's programming constructs. Most of them implement algorithms that complement the mathematical methods studied in calculus and advanced calculus.

VI.2.1 Illustrating Midpoint, Trapezoidal and Simpson's Approximations to Integrals

Problem: Find and illustrate the midpoint, trapezoidal and Simpson's approximations to $\int_0^1 e^{-5\sin^2(x)}dx$.

Solution: We present two subroutines that compute and illustrate the midpoint and trapezoidal methods respectively. Using these subroutines, we compute Simpson's approximations and compare the three approximations to CC's value for the given integral.

The first subroutine below computes the midpoint approximation, displays the graph of the function being integrated and displays rectangles corresponding to the midpoint approximations. The calling sequence for the subroutine is

`MidPt(f,r,s,n)`, where *f* is the function to be integrated, *r* and *s* are the limits of integration, and *n* is the number of intervals to use in the approximation.

The second subroutine, `Trap(f,r,s,n)`, is very similar, but produces the trapezoidal approximation to an integral and sketches the approximating trapezoids rather than approximating rectangles.

```
Function Midpoint(%f,r,s,n)
    // Purpose:  Graph f(x) and approximate its integral with  midpoint
    //            method.  Actually draw rectangles that approximate the
    //            integral.
    // Input:  f    :  name of function to integrate
    //         r,s  :  real numbers, limits of integration
    //         n  :  integer, number of intervals to use in the
    //               approximation
    // Return:  value of midpoint approximation to integral

    // First Graph f(x), automatically setting correct vertical limits
  dx = (s-r)/n                          // Width of each rectangle
  fx = list(f(x),x=r+dx/2 to s-dx/2 step dx)    // midpoint values of f(x)
  high = max(fx¦0)                      // highest point on graph or 0
  low = min(fx¦0)                       // lowest point on graph or 0
  window(r,s,low,high)                  // window to hold graph
  graphics                              // keep CC in graphics mode
  graph(f(x),x)                         // graph function to be integrated

    // Initialize variables for computing midpoint approximation
  linestyle(0)                          // use solid lines for rectangles
  xleft = r                             // left side of rectangle
  integral = 0                          // accumulate sum of heights

    // Compute components of midpoint approximation, sketching a
    // rectangle for each component.
  for j = 1 to n do
      xright = xleft + dx               // right side of rectangle
      y = fx[i]                         // height of rectangle
      rect(xleft,0,xright,y)            // draw rectangle
      integral = integral+y             // add height to sum of midpoints
      xleft = xright                    // left side of next rectangle
    end

  text                                  // hold graphics until <ENTER> pressed
  return(integral*dx)                   // midpoint approximation
end
```

You can use these function subroutines by calling them from the Input Window as you would any built-in function of CC. Remember to use the *name* of the function in the calling sequence without any variables and without the "%" sign; the "%" sign is included only in the subroutine definition. For example, to approximate $\int_0^1 e^{-5\sin^2(x)}dx$, first define the function to be integrated:

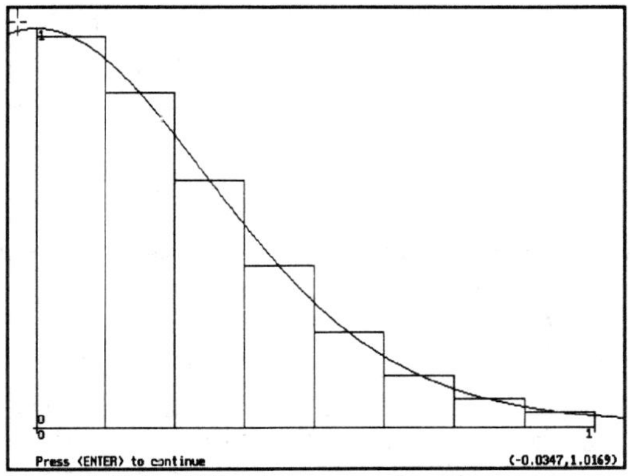

Fig. VI.2.1: Midpoint approximation to $\int f(x)dx$.

```
f(x) = exp(-5*sin(x)²)
```

Next compute and display the midpoint and trapezoidal approximations (Fig. VI.2.1 and Fig. VI.2.2):

```
mid = MidPt(f,0,1,8)
trapezoid = Trap(f,0,1,8)
```

Finally, compute Simpson's approximation, together with CC's approximation to the integral and the differences between CC's approximation and the midpoint, trapezoidal and Simpson's approximations:

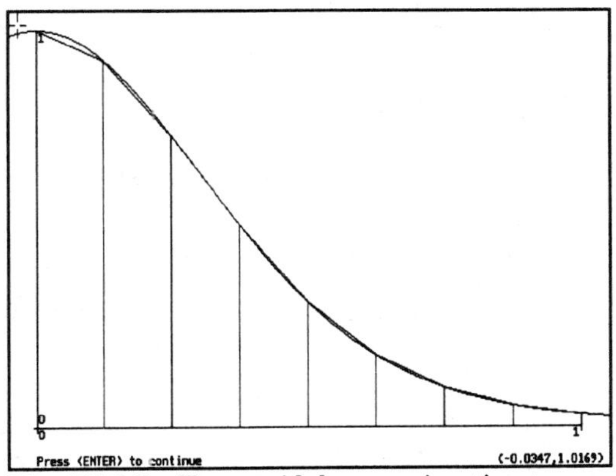

Fig. VI.2.2: Trapezoidal approximation to $\int f(x)dx$.

simpson = (2*mid+trapezoid)/3
CC = In(f(x),x=0,1)
miderror = abs(mid-CC)
traperror = abs(trapezoid-CC)

The results are:

```
MID =  0.41717515493
TRAPEZOID =  0.41692004311
SIMPSON =  0.41709011766
CC =  0.41709044091
MIDERROR =  0.000084714024524
TRAPERROR =  0.00017039779823
```

```
Function Trap(%f,r,s,n)
    // Purpose:  Graph f(x) and approximate its integral with  trapezoidal
    //           method.  Draw trapezoids that approximate the integral.
    // Input:  f   :  name of function to integrate
    //         r,s  :  real numbers, limits of integration
    //         n   :  integer, number of intervals to use in the
    //                approximation
    // Return:  value of trapezoidal approximation to integral

    // First Graph f(x), automatically setting correct vertical limits
  dx = (s-r)/n                         // Width of each trapezoid
  fx = list(f(x),x=r to s step dx)     // values of f(x) at ends of
                                       // trapezoids
  high = max(fx¦0)                     // highest point on graph or 0
  low = min(fx¦0)                      // lowest point on graph or 0
  window(r,s,low,high)                 // window to hold graph
  graphics                             // keep CC in graphics mode
  graph(f(x),x)                        // graph function to be integrated

    // Initialize variables for computing midpoint approximation
  linestyle(0)                         // use solid lines for rectangles
  xleft = r                            // left edge of trapezoid
  integral = fx[1]/2-fx[n+1]/2         // will add fx[2]+...+fx[n+1] to this

    // Compute components of midpoint approximation, sketching a rectangle
    // for each component.
  for j = 2 to n+1 do
      xright = xleft + dx              // right edge of rectangle
      yleft = fx[i-1]                  // left side of trapezoid
      yright = fx[i]                   // right side of trapezoid
      line(xleft,0,xleft,yleft)        // draw trapezoid : left side
      line(xleft,yleft,xright,yright)      // Top side
      line(xright,yright,xright,0)     // right side
      integral = integral+yright       // add right side to sum of sides
      xleft = xright                   // left edge of next rectangle
    end

  text                                 // hold graphics until <ENTER> pressed
 return(integral*dx)                   // midpoint approximation
end
```

You can see that the midpoint approximation is about twice as good as the trapezoidal approximation. We cannot compute the error in Simpson's approximation without more theoretical considerations, since CC's own computation of the integral is itself based on Simpson's approximation.

VI.2.2 Illustrating Secant and Newton's Methods for Solving Equations

Problem: Approximate the real root of the polynomial $x^3 + 9$ using the secant and Newton's methods. Illustrate the solutions graphically.

Solution: We are going to present subroutines that will graph a given curve $y = f(x)$ on a domain $[a, b]$ and the secant or tangent lines that are used to solve the equation $f(x) = 0$. We begin with the secant method.

The subroutine `Secant` starts by determining if there is a root in $[a, b]$ by checking that the values $f(a)$ and $f(b)$ have opposite signs. If a root exists, the subroutine graphs $y = f(x)$ on the interval $[a, b]$. It then draws a line from $(a, f(a))$ to $(b, f(b))$ and determines the point g in the interval $[a, b]$ where the line crosses the x-axis. Next the subroutine decides if the desired root is to the left or right of g, and repeats the process on either the interval $[a, g]$ or $[g, b]$, whichever contains the root. The procedure stops when $|f(g)| < eps$, and the last value of g is returned as the approximate solution to $f(x) = 0$. Each line drawn to find a better approximation to the root of $f(x)$ is shown on the screen.

To use this subroutine to solve the equation $x^3 + 0.9 = 0$, enter the following lines:

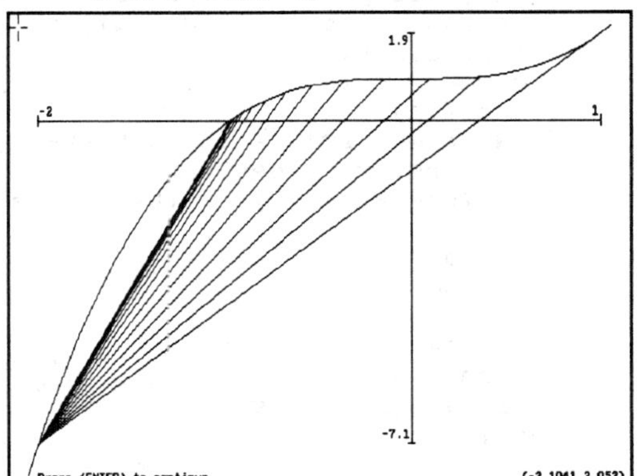

Fig. VI.2.3: Solving $x^3 + 0.9 = 0$ with the secant method.

```
function Secant(%f,a,b,eps)
  // Purpose:  solve f(x) = 0 by secant method in the interval [a,b].
  //           Return solution x when abs(f(x)) < eps.  Also graph f(x)
  //           and the secant lines used to find the solution.
  // Input:  f   :  Name of function to use
  //         a,b  :  real, boundaries of interval containing solution
  //         eps  :  positive real, error tolerance for solution
  // Return:  approximate solution x to equation f(x) = 0

    // values of f(x) at endpoints of interval [a,b] containing solution
  c = f(a)
  d = f(b)
    // if c and d have same sign, possibly no root in [a,b]
  if (c > 0 and d > 0) or (c < 0 and d < 0)
      write('Possibly no root in  [',a,',',b,]')
      exit
    end
    // Define window to display curve and graph f(x)
  if c < d
      window(a,b,c,d)
    else
      window(a,b,d,c)
    end
  graphics                              // turn on graphics mode
  graph(f(x),x)                         // sketch function to be solved
  LineStyle(0)                          // Draw all lines as solid lines
    // Repeatedly replace [a,b] by a smaller interval containing a
    // solution to f(x) = 0, until a solution satisfying
    // abs(f(x)) < eps is found
  Repeat
    // At this point, there is a root in [a,b].  Compute point g where
    // secant line from (a,f(a)) to (b,f(b)) crosses the x-axis.
      line(a,c,b,d)                     // draw secant line (a,f(a))-(b,f(b))
      g = a - c*(b-a)/(d-c)             // where secant line crosses x axis
      h = f(g)
    // Next we replace the interval [a,b] with either [a,g] or [g,b],
    // whichever is certain to contain the root.
      if (h>0 and d<0) or (h<0 or d>0)      // if root in [g,b]
          b=g
          d=h
        else                            // else root is in [a,g]
          a=g
          c=h
        end
    Until abs(h) < eps                  // continue until h = f(g) is small
  return(g)                             // best solution to f(x) = 0
  text                                  // hold graphics until <ENTER> pressed
end
```

```
f(x) = x^3+.9
a=secant(f,-2,1,.01)
a1 = f(a)
```

CC will display the graph shown in Fig. VI.2.3 and return the value a = -0.962114. When you evaluate $f(a)$, the result will be $a1$ = 0.0094058, which is smaller (in absolute value) than eps = 0.01.

To solve the same equation using Newton's method, we introduce a subroutine `Newton(f,a,eps)` that calculates roots using Newton's method and illustrates the calculation by graphing the function and the tangent lines used to find the root. `Newton(f,a,eps)` starts by graphing the function $f(x)$ in the window currently in place. Next, beginning with $x = a$, it computes a better approximation $b = a - f(a)/f'(a)$ to the root of $f(x)$ and draws the tangent line from $(a, f(a))$ to $(b, 0)$ and the vertical line from $(b, 0)$ to $(b, f(b))$. Then, replacing a with b, the process is repeated until an approximation a is created such that $|f(a)| < eps$. It is this value of a that is returned by `Newton`.

To solve the equation $x^3 + 0.9 = 0$ with `Newton`, enter the following lines:

```
f(x) = x^3+.9
b = Newton(f,1,0.01)
b1 = f(b)
```

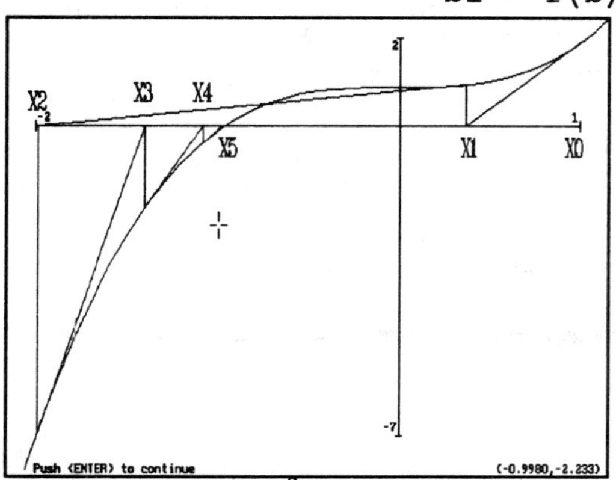

Fig. VI.2.4: Solving x^3 + 0.9 = 0 by Newton's method.

The graph shown in Fig. VI.2.4 will be displayed, and the value b = -0.965662 will be returned. Labels have been added to the graph showing the first few approximate solutions to $x^3 + 0.9 = 0$ produced by Newton. When you evaluate $f(b)$, the result will be $b1$ = -0.000481856 which is smaller (in absolute value) than eps = 0.01.

Both these subroutines return roots which are approximate zeros of

$x^3 + 0.9$. A more accurate approximation can be found by using CC's built-in equation solver:

```
solve(f(c)=0,c=-1)
```

which returns c = -0.965489

```
function Newton(%f,a,eps)
 // Purpose:  use Newton's method to solve f(x)=0 starting at  x = a.
 //           Also graph the tangent lines used to find the root.
 //           This subroutine uses whatever window is present when it
 //           starts.
 // Input:  f   :  name of function to solve
 //         a   :  real, initial guess for root
 //         eps  :  positive real, tolerance for solution
 // Return:  approximate solution to f(x) = 0

 erase              // start with a clean screen
 graphics           // get into graphics mode
 graph(f(x),x)     // graph the function to be solved

 Repeat
   // replace the current best estimate a for the root
   // with b = a - f(a)/f'(a)
     b = a - f(a)/dif(f(x),x=a)
     Line(a,f(a),b,0)                 // Draw tangent line (a,f(a)) to (b,0)
     Line(b,0,b,f(b)                  // Vertical line (b,0) to (b,f(b))
     a = b                            // New best estimate for root

   // stop if last approximation undefined (because
   // f'(a) = 0) or if f(a) small enough.
   until a == undefined or abs(f(a)) < eps

 return(a)   // value of subroutine is best guess at root
 text        // hold graphics until user presses <Enter>
end
```

VI.2.3 Differential Equations

Problem: Compare the efficiency of the Euler and the predictor-corrector algorithms for solving the initial value problem:

$$\begin{aligned} y' &= Sin(5t) + y \\ y(0) &= 1 \end{aligned}$$

and approximating the solution $y(2)$. Your algorithm should graph its solution on the screen.

Solution: This problem has a solution in closed form, which is:

$$y(t) = \frac{-1}{26}\sin(5t) - \frac{5}{26}\cos(5t) + \frac{31}{26}e^t$$

This exact solution will measure the accuracy of our approximate solution. It is entered into CC as:

```
y(t) = (-1/26)sin(5t)-(5/26)cos(5t)+(31/26)exp(t)
```

We will approximate the value of $y(2)$, which according to CC is 8.992312224.

Here is the general method for approximating the solution $y(x)$ to an initial value problem:

$$\begin{aligned} y' &= f(t,y) \\ y(a) &= b \end{aligned}$$

1. divide the interval $[a, x]$ into n subintervals of length dt;

2. starting with $b = f(a)$, successively compute approximations for $y(t)$ at $t = a+dt$, $a+2dt$, ..., x.

The main difficulty in approximating the solution to a differential equation is approximating the slope between the (previously approximated) solution point $(t, y(t))$ and the next point $(t+dt, y(t+dt))$. Call this value *slope*. Once *slope* is approximated, you can compute $y(t+dt) = y(t) + slope*dt$.

Euler's method uses the approximation $slope = f(t, y(t))$, since that would be the slope of a solution curve $y(t)$ passing through the point $(t, y(t))$. The subroutine

`Euler(f,a,b,x,n)` implements this algorithm. To test it, enter the following commands:

```
Window(0,2,0,10)
f(t,y) = Sin(5t)+y
a1=Euler(f,0,1,2,10)
a2=Euler(f,0,1,2,100)
```

CC returns *a1* = 7.2894798865 and *a2* = 8.7942343622, a considerable improvement. CC also displays the graph of y(t) on the interval [0, 2] as the approximation y(2) is being computed.

A better algorithm is the predictor-corrector method. As with Euler's method, the predictor-corrector method computes a series of approximations for $y(t)$ at $t = a+dt, a+2dt, ..., a+n{\cdot}dt = x$. Assuming $y(t)$ is already computed, the computation of $y(t+dt)$ is accomplished in five steps:

```
Function Euler(%f,a,b,x,n)
  // Purpose:  use n-step Euler method to compute y(x) if y'(t) = f(t,y)
and
  // y(a) = b.  Also, graph the solution y(t) on the interval [a,b]
  // Input:   f : name of function defining first order ODE
  //          a,b : real, y(a) = b is initial condition
  //          x : real, algorithm computes y(x)
  //          n : positive integer, number of intervals to use in
  //              approximation
  // Return:  Euler's approximation to y(x)

  graphics                              // Put CC in graphics mode
  dt = (x-a)/n                          // increment of t for each step
  y = b                                 // g(a)

    // Repeat y(t+dt) = y(t) + slope*dt, t=t+dt until t reaches x
    // and y reaches y(x)
  for t = a to x - dt step dt do
      oldy = y                          // approximate value of y(t)
      y = y + f(t,y)*dt                 // Approximate value of y(t+dt)
      line(t,oldy,t+dt,y                // extend graph of y(t)
    end

    // at this point, y is approximate value of y(x)
  text                                  // end graphics mode
  return(y)
end
```

1. the approximate slope $y'(t)$, $slope1 = f(t,y(t))$, is computed;
2. the predicted value for $y(t+dt)$, $ytemp = y(t)+slope1 \cdot dt$, is computed, as in the Euler method;
3. the approximate slope $y'(t+dt)$, $slope2 = f(t,ytemp)$, is computed;
4. a corrected approximation for the slope of $y(t)$ on $[t, t+dt]$, $slope = (slope1+slope2)/2$, is computed;
5. finally, an improved approximation for $y(t+dt)$, $y(t)+slope \cdot dt$, is computed.

The subroutine PC implements the predictor-corrector method. To test it, enter the formulas:

```
Function PC(%f,a,b,x,n)
  // Purpose:  use n-step predictor-corrector method to compute y(x) if
  //           y'(t) = f(t,y) and y(a) = b.  Also, graph the solution y(t)

  //           on the interval [a,b]
  // Input:   f : name of function defining first order ODE
  //          a,b : real, y(a) = b is initial condition
  //          x : real, algorithm computes y(x)
  //          n : positive integer, number of intervals to use in
  //              approximation
  // Return:  predictor-corrector approximation to y(x)

  graphics                          // Put CC in graphics mode
  dt = (x-a)/n                      // increment of t for each step
  y = b                             // g(a)

    // Repeat y(t+dt) = y(t) + slope*dt, t=t+dt until t reaches x
    // and y reaches y(x)
  for t = a to x - dt step dt do
      oldy = y                      // approximate value of y(t)
      slope1 = f(t,y)               // first approximation to slope
      ytemp = y + slope*dt          // first approximation to y(t+dt)
      slope = (slope1 + f(t+dt,ytemp))/2  // better approximation to slope
      y = y + slope*dt              // Approximate value of y(t+dt)
      line(t,oldy,t+dt,y)           // extend graph of y(t)
    end

    // at this point, y is approximate value of y(x)
  text                              // end graphics mode
  return(y)
end
```

```
Window(0,2,0,10)
f(t,y) = Sin(5t)+y
b1 = PC(f,0,1,2,10)
b2 = PC(f,0,1,2,100)
```

CC returns $b1$ = 8.7440259344 and $b2$ = 8.992312224. The accuracy of the predictor-corrector method with 10 steps is almost equal to the accuracy of Euler's algorithm with 100 steps.

Fig. VI.2.5 displays, from bottom to top, the graphs produced by `Euler(f,0,1,2,10)` and `PC(f,0,1,2,10)` and the graph of the exact solution $y(t)$.

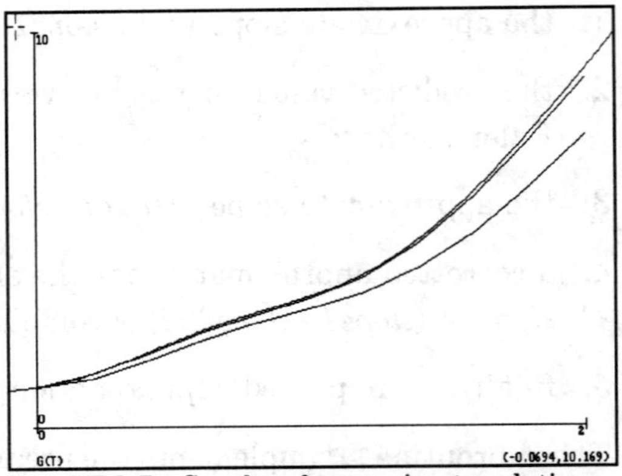

Fig. VI.2.5: Graphs of approximate solutions to $y'=y+\sin(t)$, $y(0)=1$.

VI.2.4 Graphing f(x,y) = 0

Problem: Graph the equations $x^2 + y^2 = 1$ and $x^3 + x^2 = y^2$.

Solution: The subroutine `GraphEqn(f,a,b,n)` graphs the level curve $f(x,y) = 0$ by using the `Solve` operator to move the initial point (a, b) to the curve, then constructing a series of line segments approximating the curve. Each line segment begins at a point (x, y) on the curve and ends at another point $(x1, y1)$ on the curve. The point $(x1, y1)$ is computed from (x, y) by following the tangent vector to $f(x, y)$ beginning at (x, y) to a point near the curve that is then corrected to $(x1, y1)$ on the curve.

One complication has been added to allow the algorithm to trace a curve through a singularity. When a branch of a curve crosses a singularity, its tangent vector may reverse direction. The subroutine senses this reversal and reverses the tangent vector back to its original direction along the curve so that drawing can continue in the original direction.

In Fig. VI.2.6, `GraphEqn` has been applied to two equations, the circle $x^2 + y^2 = 1$ and the node $x^3 + x^2 = y^2$, by entering the lines:

```
Window(-1.33,1.33,-1,1)
f(x,y) = x^2+y^2-1
GraphEqn(f,-1,0,60)
f(x,y) = x^3+x^2-y^2
GraphEqn(f,-1,0,100)
```

In tracing the second equation through its node, the algorithm passes through a singularity at the origin. You will hear the computer beep when this happens.

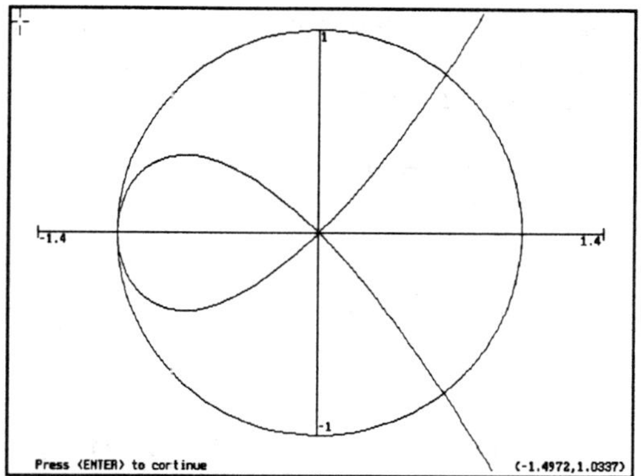

Fig. VI.2.6: Graphs of $x^2+y^2 = 1$ and $x^3+x^2 = y^2$.

```
Function OnScreen(x,y)
  // Purpose:  determine if (x,y) is within the present graphing window
  // Input:  x,y : real, point to check
  // Result:  true if (x,y) in the graphing window, false otherwise
  Return((x >= WindowLeft and x <= WindowRight
             and y >= WindowBottom and y <= WindowTop)
end

Procedure GraphEqn(%f,a,b,n)
  // Purpose:  Graph level curve of f(x,y) = 0 beginning at point nearest
  //           (a,b).  Use n steps each way from (a,b), each step is about
  //           2% of the screen size.
  // Input:  f : name of function to graph
  //         a,b : real, start graphing at point on curve nearest (a,b)
  //         n : positive integer, number of steps to take each way from
  //            starting point
  graphics                          // Put CC into graphics mode
    // Move (a,b) vertically or horizontally to nearest point on curve.
    // Move vertically if tangent to curve is more horizontal than vertical,
    // or if gradient vector to f(x,y), (df/dx, df/dy), more vertical than
    // horizontal, i.e. if abs(df/dx) <abs(df/dy)
  if abs(dif(f(x,b),x=a)) < abs(dif(f(a,y),y=b))
      Solve(f(a,b) = 0, b = b)      // move b to solution (a,b)
    else
      solve(f(a,b) = 0, a = a)      // move a to solution (a,b)
    end
```

```
width = windowright - windowleft  // width of window
height = windowtop - windowbottom // height of window
for k = 1 to 2                    // draw curve in two directions from
                                  // (a,b)
    steps = 0                     // Count number of steps
    x = a                         // (x,y) will be point on curve
    y = b                         // start at (x,y) = (a,b)
    savev1 = 0                    // will store components of tangent
    savev2 = 0                    // vector
    f(x,y) = eval(-f(x,y))        // change direction to draw from prior
                                  // direction.  Eval avoids infinite
                                  // recursion.
    f1(x,y) = eval(dif(f(x,y),x)) // f1 = df/dx
    f2(x,y) = eval(dif(f(x,y),y)) // f2 = df/dy
    while steps < n and Onscreen(x,y)    // draw one branch of curve until
                                         // off-screen or taken n steps
        v1 = -f2(x,y)             // (v1, v2) perpendicular to gradiant of
        v2 = f1(x,y)              // f(x,y), so tangent to f(x,y) = 0
              // If (v1,v2) points in opposite direction from previous
              // tangent vector, then we just passed a singularity of the
              // curve, so we will reverse the tangent vector to continue
              // drawing the curve in the same direction.
        if (v1*savev1 < 0) and (v2*savev2) < 0      // if direction changed
            beep                  // sound indicates that direction
                                  // reversed
            f1(x,y) = eval(-f1(x,y))     // reverse direction of partial
            f2(x,y) = eval(-f2(x,y))     // derivatives
            v1 = -v1              // and reverse last tangent vector
            v2 = -v2
          end
              // draw next step of curve
        q = 50*max((abs(v1/width), abs(v2/height))) // each step will have
                                                     // length about 1/50th
                                                     // size of screen
        x1 = x+v1/q               // (x1,y1) approximately on curve
        y1 = y+v2/q
        if abs(v1) > abs(v2)      // if tan vector more horizontal
            FSolve(f(x1,y1) = 0, y1 = y1)     // correct y1
          else
            FSolve(f(x1,y1) = 0, x1 = x1)     // correct x1
          end
        line(x,y,x1,y1)           // extend graph of curve
        x = x1                    // update point (x,y) on curve
        y = y1
        steps = steps + 1         // and count of steps taken
```

```
            savev1 = v1                 // and last tangent vector
            savev2 = v2
          end                           // While onscreen and steps < 50
      end                               // drawing one direction
    text                                // cancel graphics mode
end
```

VI.2.5 Vector Fields and Level Curves

Problem: Sketch the dipole electric field for a positive charge at (1,0) and a negative charge at (-1,0) and draw some level curves through the field.

Solution: The mathematical expression for the desired vector field is:

$$f(x,y) = \frac{(1-x,-y)}{((1-x)^2+y^2)^{3/2}} - \frac{(-1-x),-y)}{((1-x)^2+y^2)^{3/2}}$$

Note that $f(x, y)$ is a *vector-valued function* which is the difference of two vector fields, one directed away from (-1,0) and the other towards (1,0) whose strengths are inversely proportional to their distances from their centers. Thus $f(x, y)$ is the desired vector field.

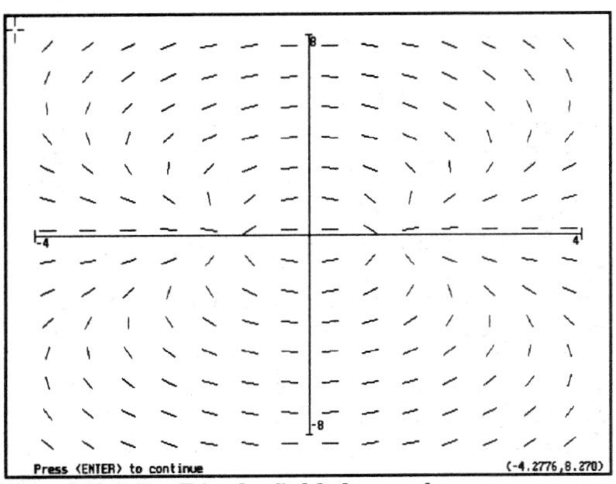

Fig. VI.2.7: Dipole field drawn by VectorField.

To sketch the vector field, we will construct a subroutine `VectorField` that samples it at an $n \times n$ grid in the window and draws short line segments at the grid points parallel to the vector field. The only difficulty is making all the line segments the same physical length on the screen no matter how the screen is scaled. The subroutine solves that problem.

You can display the dipole field (Fig. VI.2.7) by entering the lines:

```
// dipole field with poles at (+/-1,0).
// Field points away from (-1,0) and towards (1,0)
f(x,y)=(1-x,-y)/((1-x)²+(y)²)^(3/2)
        - (-1-x,-y)/((-1-x)²+(y)²)^(3/2)
Window(-4,4,-8,8)
vectorfield(f,15)
```

To draw a level curve through this vector field, use the subroutine `LevelC`, which is similar to the one used to graph implicit functions (IV.2.4). Fig. VI.2.8 shows four level curves through the dipole field, created by the commands:

```
procedure vectorfield(%f,n)
  // Purpose:  Draw n x n array of line segments parallel to vector field
  //           f(x,y)
  // Input:  f : name of function defining vector field
  //         n : positive integer, size of array of direction lines to
  //             display
  graphics                        // start graphics mode
  width = windowright-windowleft  // window width
  height = windowtop-windowbottom // window height
  dx = width/n                    // horizontal distance between marks
  dy = height/n                   // vertical distance
  w = (width,height)/(5n)         // factor used below
  for x = windowleft+dx to windowright-dx step dx     // draw n x n array
      for y = windowbottom+dy to windowtop-dy step dy // of line segments
                                                      // parallel to field
          v = f(x,y)              // vector at (x,y)
          s = sqrt(sum((v/w)^2))  // This is tricky.  s is the length of
                                  // the vector v/w, where v/w is a
                                  // vector with components proportional
                                  // to the components of v divided by
                                  // the size of the corresponding
                                  // window dimension.  s is large if v
                                  // is large relative to the window
          line(x-v[1]/s,y-v[2]/s,x+v[1]/s,y+v[2]/s)   // parallel to v
      end
   end
  text                            // hold graphic until <ENTER> pressed
end
```

```
Levelc(f,-1.015,-.05,.1)
LevelC(f,-1.015,0.1,.1)
LevelC(f,-1.015,0.02,.1)
LevelC(f,-1.015,-0.015,.1)
```

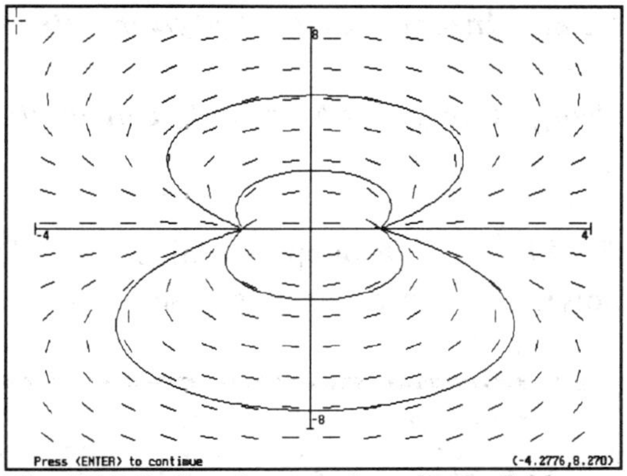

Fig. VI.2.8: Level curves in a dipole field.

```
Procedure LevelC(%f, a, b, dt)
  // Purpose:  f(x,y) is a vector field.  Beginning at (a,b), draw level
  //           curve until offscreen or 200 steps taken.
  // Input:  f  :  name of vector-valued function defining level curve
  //         a,b  :  real, (a,b) is the point where the level curve starts
  //         dt  :  real, the length of each step of the curve

  graphics
  i = 0
      // while onscreen and less than 200 steps taken, draw a line segment
      // from (a,b) in the direction of the field f(x,y).
  while i < 200 and a <> undefined and b <> undefined
     and a > windowleft and a < windowright
     and b > windowbottom and b < windowtop
      i = i+1                        // increment number of line segments
      a1=a                           // drawn, save current point on curve,

      b1=b                           // advance (a,b) to next point
      v=f(a,b)                       // direction vector at starting point
      s = sqrt(sum(v^2))             // length of v
      a=a+v[1]/s*dt                  // (a,b) is distance dt in direction
      b=b+v[2]/s*dt                  // v from (a1,b1)
      Line(a1,b1,a,b)                // draw line segment from old point to
                                     // new point
    end
  text                               // hold graphic until <ENTER> pressed
end
```

Chapter 6
Applications

VI.2.6 Interpolating Polynomials

George Corliss, of Marquette University, developed this application. My thanks to him.

Problem: Develop an algorithm for computing the Newton interpolating polynomial on an arbitrary set of points.

```
//  Workspace:   NEWTON1.CC3
//  Author:      Dr. George F. Corliss, Marquette University
//  Written:     14-FEB-1990
//  Purpose:     Illustrate Newton form of an interpolating
//               polynomial
//  Contents:    Get_x, Get_y, Div_Diff, Newton

Function Div_Diff(x, y)
  // Purpose:  Compute a divided difference table
  // Input: x   :  list of reals, X coordinates to interpolate
  //        y   :  list of reals, Y coordinates to interpolate
  // Output: list of reals Div_Diff, divided difference table
  a = y
  n = length(a)
  for i = 2 to n
      for j = i to n
          ni = n - j + i
          a[ni] = (a[ni] - a[ni-1]) / (x[ni] - x[n-j+1])
        end                         // for j
    end                             // for i
  return(a)                         // Divided difference table
end                                 // Div_Diff

Function Newton(x, a, t)
  // Purpose:  Evaluate the Newton form of an interpolating polynomial.
  // Input:   x   :  list of real, X coordinates to interpolate
  //          a   :  list of real, divided differences from Div_Diff
  //          t   :  real, point at which to evaluate polynomial
  // Output: real, Newton value of interpolating polynomial
  n = length(a)
  f = a[n]
  for i = 1 to n-1
      f = f * (t - x[n-i]) + a[n-i]
    end                             // for i
  return (f)                        // value of polynomial
end                                 // Newton
```

Solution: We shall present a routine that computes the Newton interpolating polynomial on a given set of points and also a game based on this algorithm. The game requires the player to select a set of points so that the Newton Polynomial through them approximates a given curve. The degree of approximation is measured by computing the integral of the square difference between the given curve and the interpolating polynomial in the region displayed on the screen.

The algorithm that computes the values of the interpolating polynomial is divided into two parts. The subroutine `Div_Diff(x,y)` takes two lists or arrays of values (of the same length) x and y, and returns an array containing the divided difference table for these points. The second subroutine, `Newton(x,a,t)`, accepts as input the list x of x-values, the table a of divided differences produced by `Div_Diff`, and a value t, and returns the value of the interpolating polynomial at t. By dividing the algorithm into two parts, you can compute the divided difference table once for the given points, and then find as many values as you need of the interpolating polynomial.

I have turned Dr. Corliss' algorithms into a game with two more subroutines. The first one, `Points(n)`, allows the user to select n points on the screen by moving the crosshairs to them and pressing <Enter> one at a time. The points will be marked with crosses, and the subroutine will return a list (x, y) of lists; x contains the x-coordinates of the chosen points, and y contains the y-coordinates. The second subroutine, `NewtonInterp(%f,a,b,c,d,n)`, actually controls the game. It creates a graphing window $a \leq x \leq b$, $c \leq y \leq d$ and graphs the function f(x) in the window. The user then selects n points in the window determining an interpolating polynomial of degree n-1. The object is to pick points so that the polynomial approximates the given curve as closely as possible. The subroutine then graphs the interpolating polynomial, so you can see how closely it approximates the given curve, and returns the integral of the square difference between the two functions. The object of the game is to make this square difference as small as possible.

We can initiate the game with the commmands:

```
function Points(n)
  // Author:  D. Meredith
  // Date 2/22/90
  // Purpose:  Mark n points on screen and return coordinates in lists
  //           x and y.  Assumes already in graphics mode
  // input: n  :  positive integer, number of points to be marked
  // output : a doubly indexed list a[1..2,1..n] of the x- and
  //          y-coordinates of the marked points.  a[1] is list
  //          of x-coordinates; a[2] is list of y-coordinates
  //
  x=()                                 // empty lists
  y=()
  For i = 1 to n do
     pause                             // user moves crosshairs to next point
                                       // and presses enter
     mark(crossx,crossy)               // mark the point
     write@c(i)                        // write index number there too
     x = x¦crossx                      // append its coordinates to x and y
     y = y¦crossy
    end
  write(n,'points marked--push <Enter> to continue')
  return((x,y))                        // two list values returned as one
list
end

function NewtonInterp(%f,a,b,c,d,n)
  // Author:  D. Meredith
  // Written : 2/22/90
  // Purpose: produce and graph Newton's interpolation of y = f(x) in
  //          window (a,b)x(c,d) using n points. First display graph of
  //          f(x), let user pick n points, then compute and graph
  //          interpolating polynomial.
  // Input:   f, name of function
  //          a,b,c,d  :  real, Window limits
  //          n  :  positive integer, number of points
  // Output:  square error integrated between f(x) and interpolating
  //          polynomial in output window
  window(a,b,c,d)                      // set window
  graphics
  quickg(f(x),x)                       // graph f(x)
  p = Points(n)                        // user marks n points
  coeff = div_Diff(p[1],p[2]           // data for interpolating polynomial
  Sketch(Newton(p[1],coeff,t),t)       // graph interpolating polynomial
  text                                 // hold graphics until <ENTER> pressed
  return(fin((f(x)-Newton(p[1],coeff,x))²,x=a))
                                       // compute square error between f(x)
                                       // and interpolating polynomial
end
```

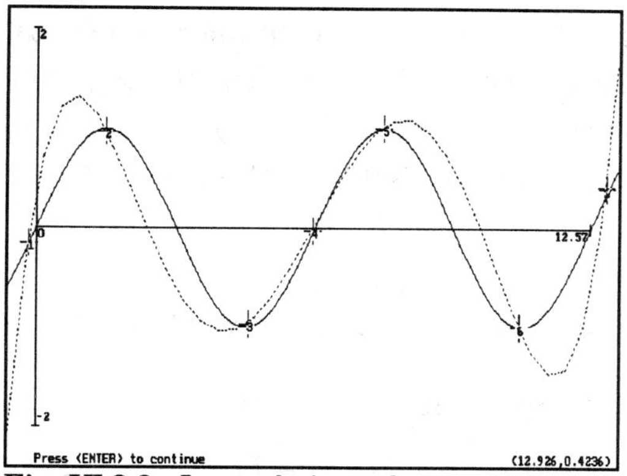

Fig. VI.2.9: Interpolating polynomial through 7 points approximating $y = \text{Sin}(x)$.

```
f(x) = sin(x)
error = NewtonInterp(f,0,
  4pi,-2,2,7)
```

Fig. VI.2.9 illustrates one play of this game. The function to be approximated is Sin(x) for $0 \leq x \leq 4\pi$, and seven points are used for the approximation. In this example, the error is 1.54. I think students could learn a lot about the interpolating polynomial by having a contest with this game.

VI.2.7 Cubic Splines

This application is due to James E. McKenna of SUNY, Fredonia. I appreciate his agreeing to have it printed here.

Problem: Find the cubic spline through a given set of points.

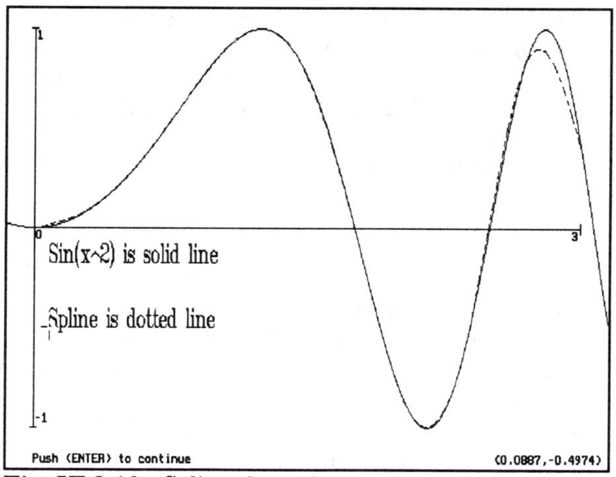

Fig. VI.2.10: Spline through 11 points on the curve $y = \text{Sin}(x^2)$.

Solution: A cubic spline passing through a given set of points is a curve pieced together from segments of cubic curves attaching each given point to the next one in such a way that, where two curves meet at a point, they have the same first and second derivatives. The subroutine `SplineCoef(x,y)` takes lists of x- and y-coordinates for the given points and returns a list of spline coefficients. (The x-coordinates must be given in increasing order.) These coefficients, together with the initial data, are used by the subroutine `SplineEval(xx, x, y, SplnCoef)` to evaluate the spline at the point xx

To test these routines, enter the lines:

```
x = list(t,t=0,3,.3)           // 0, 0.3, 0.6, ..., 3
y = sin(x^2)                   // sin(0^2), ..., sin(3^2)
window(0,3,-1,1)
graph(sin(x^2),x)              // Graph the function
splncoef = splinecoef(x,y)
Paramg(xx,SplineEval(xx,x,y,splncoef),xx = 0 to 3 step 60)
                             // graph the spline approximation
Write@c('Sin(x^2) is solid line',2)     // Label graphs
Write@c('Spline is dotted line',2)
```

```
Function SplineCoef(x,y)
  // Purpose:  Computes the coefficients of the natural cubic spline
  //           corresponding to the points whose abcissas are in the list x
  //           and whose ordinates are in the list y.
  // Input:  x,y -- list of real, x- and y- coordinates for points on spline.
  //         x is assumed ordered in increasing order
  // Output:  list of real, second derivates of spline at given points
  n = length(x)
  h = list(0, j=1 to n)          // create list before filling it
  b = h
  u = h
  v = h
  z = h
  for j = 1 to n-1
      h[j] = x[j+1] - x[j]// dx[j]
      b[j] = (y[j+1] - y[j])/h[j]// dy/dx[j]
    end
  u[2] = 2*(h[1] + h[2])
  v[2] = 6*(b[2] - b[1])
  for j = 3 to n-1
      u[j] = 2*(h[j] + h[j-1]) - h[j-1]²/u[j-1]
      v[j] = 6*(b[j] - b[j-1]) - h[j-1]*v[j-1]/u[j-1]
    end
```

```
  for j = n-1 to 2 step -1
      z[j] = (v[j] - h[j]*z[j+1])/u[j]
    end
  return(z)                           // list of spline coefficients
end


Function SplineEval(xx,x,y,SplnCoef)
  // Purpose:  Evaluate at xx the spline corresponding to the points in the
  //           lists x and y whose spline coefficients are in the list
  //           SplnCoef
  // Input:  xx -- real, point to evaluate spline at
  //         x,y -- lists of real, points spline built on
  //                x is assumed ordered in increasing order
  //         SplnCoef -- list of real, spline coefficients from SplineCoef
  // Output:  real, value of spline at xx
  n = Length(x)
  if xx < x[1]                        // If xx below first spline point
      Write('WARNING--',xx,'is below domain of spline')
      p = y[1]
    else if xx > x[n]                 // If xx above last spline point
      p = y[n]
      write('WARNING--',xx,'is above domain of spline')
    else
      j = n-1
      diff = xx - x[j]
      while j >= 2 and diff < 0
          j = j-1
          diff = xx - x[j]
        end // x[j] <= xx < x[j+1]
      h = x[j+1] - x[j]
      b = (y[j+1] - y[j])/h - h*(SplnCoef[j+1] + 2*SplnCoef[j])/6
      p = 0.5*SplnCoef[j] + diff*(SplnCoef[j+1] - SplnCoef[j])/(6*h)
      p = b + diff*p
      p = y[j] + diff*p
    end
  return(p)                           // value of spline at xx
end
```

Chapter 6
Applications

VI.2.8 Chaos Theory

Problem: In the 1970's, the mathematician E. O. May investigated a simple model of population growth: if x is the population in one generation, then $f(x) = rx(1\text{-}x)$ is the population in the next generation. The parameter r represents the environment; it can vary from 0 to 4. The population x is always a fraction (of the maximum population) and varies between 0 and 1. What does this model predict about populations under different environments?

Solution: A fascinating discussion of this model, with some technical references, can be found in Gleick, *Chaos*. Viking, New York, 1987, pp. 57-81. We will present two subroutines that illustrate the model. First we plot the values of x over many generations with the subroutine `Chaos1(r,n,x)`. The parameter r describes the environment; n the number of generations plotted; and x the size of the first generation. The output from this subroutine is illustrated below.

The sequence of generations can also be superimposed on the graph of $f(x)$. The subroutine `Chaos2(r,m,n,x)` first sketches the graphs of $y = f(x)$ and $y = x$. Next it advances the sequence $x = f(x)$ for m terms, beginning with the given value of x. Then it advances the sequence $x = f(x)$ for n terms, each time drawing line

```
Procedure Chaos1(r,n,x0)
  // Purpose:  graph n values of the sequence x <- r*x*(1-x) beginning
with
  //              x = x0
  // Input:  r   :  real, coefficient of generating function
  //         n   :  positive integer, number of times to iterate
  //         x0 == real, starting point for sequence

  f(x) = r*x*(1-x)                    // generating function for sequence
  window(0,n*1.1,0,1)                 // graphing window
  graphics
  for j = 1 to n do
    oldx = x                          // last generation
    x = f(x)                          // next generation
    line(j-1,oldx,j,x)                // line between them
   end
 write@(n,x,x)                        // write size of last generation
 text
end
```

segments from (x, x) to $(x, f(x))$ to $(f(x), f(x))$. This picture shows how the graph of $f(x)$ controls the progress of the generations x. By skipping the first few generations, Chaos2 can more clearly illustrate the limiting behavior of the model.

May's simple model surprised the scientific world by predicting sequences of population values that did not converge to a stable size. For some values of r, the resulting populations oscillated between two values, for others they cycled around four or eight values, and for some values the population sizes seemed to vary randomly. For $1 < r < 3$ (Fig. VI.2.11, Fig. VI.2.12) the population approaches a stable value equal to $(r-1)/r$; for $3 < r < 3.6$ (approximately) (Fig. VI.2.13 - Fig. VI.2.18) the population approaches a periodic sequence whose length is always a power of 2; and for $r > 3.6$ (Fig. VI.2.19, Fig. VI.2.20) the population varies chaotically.

```
Procedure Chaos2(r,m,n,x)
  // Purpose:  graph n values of the sequence x = r*x*(1-x) on graph of
  //           y =  rx(1-x).  Begin with given value of x and generate m
  //           values before displaying any.  Thus display m+1'st to
m+n'th
  //           value of x.
  // Input:  r   :  real, coefficient of generating function
  //        m,n  :  positive integers, generate m hidden point, then
display
  //                 the next n values
  //         x   :  real, starting point for sequence

  f(x) = r*x*(1-x)                     // generating function for sequence
  window(0,1,0,1)
  graphics
  quickg(f(x),x)                       // graph y = f(x)
  sketch(x,x)                          // graph y = x
  for j = 1 to m do                    // advance sequence of x's for m terms
      x = f(x)
    end
 for j = 1 to n do                     // graph next n x's
      y = f(x)
      line(x,x,x,y)
      line(x,y,y,y)
      x = y
    end
 text
end
```

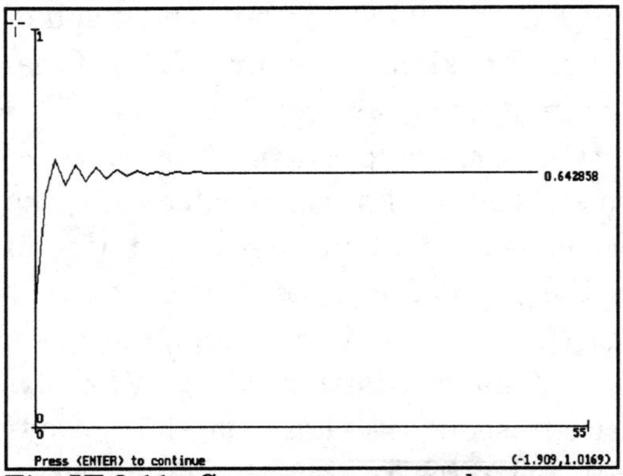

Fig. VI.2.11: Convergence to steady state when r = 2.8: `Chaos1(2.8,50,0.6)`

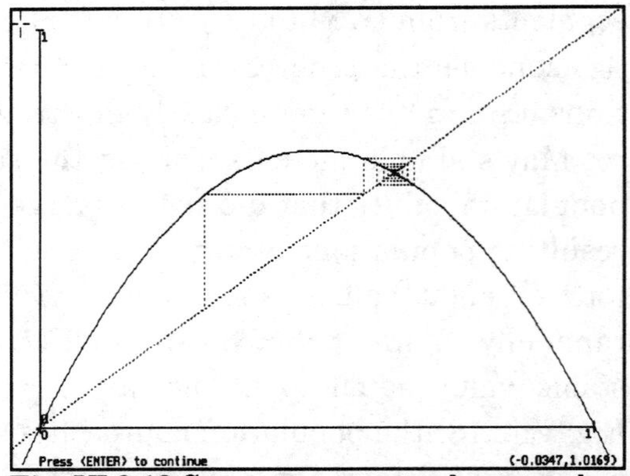

Fig. VI.2.12:Convergence to steady state when r = 2.8: `Chaos2(2.8,0,30,0.3)`

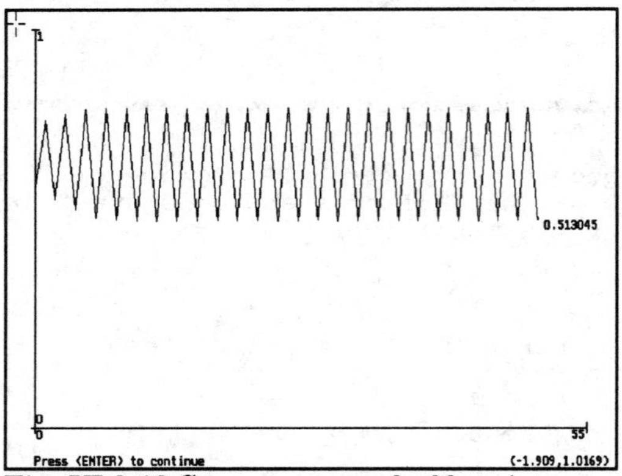

Fig. VI.2.13:Convergence to double point when r = 3.2: `Chaos1(3.2,50,0.6)`

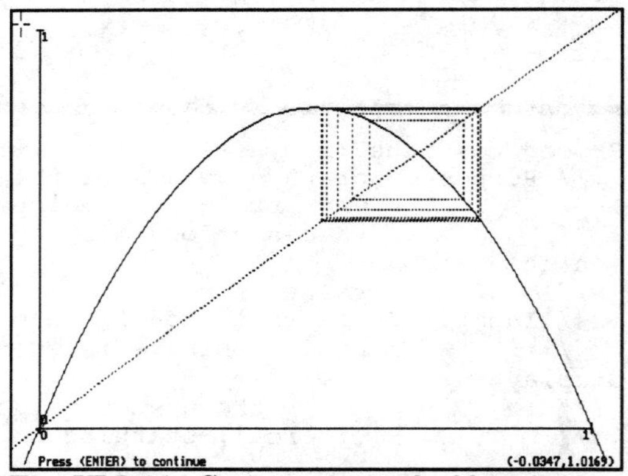

Fig. VI.2.14: Convergence to double point when r = 3.2: `Chaos2(3.2,0,30,0.6)`

A comprehensive view of the behavior of May's model can be constructed by graphing population levels x for each value of r. The subroutine `Bifurc(a,b,c,d,n,m,q)` creates this plot as follows.

1. It defines a window with $a \le r \le b$ on the horizontal axis and $c \le x \le d$ on the vertical axis;

2. Starting with $r = a$ and incrementing r by an amount $dr = (b\text{-}a)/q$ until $r = b$, it plots a column of x values above each value of r;

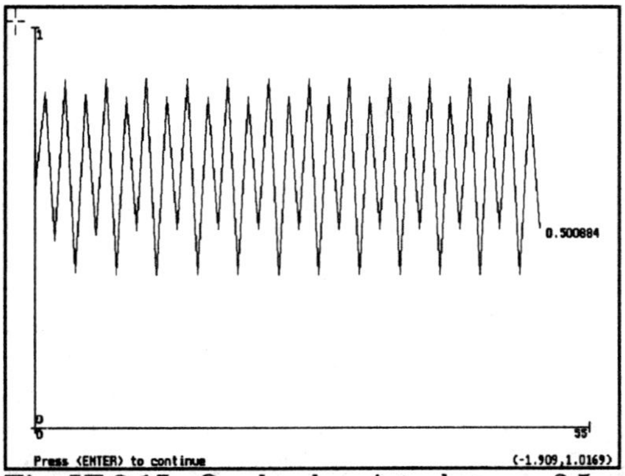

Fig. VI.2.15: Quadruple point when r = 3.5:
`Chaos1(3.5,50,0.6)`

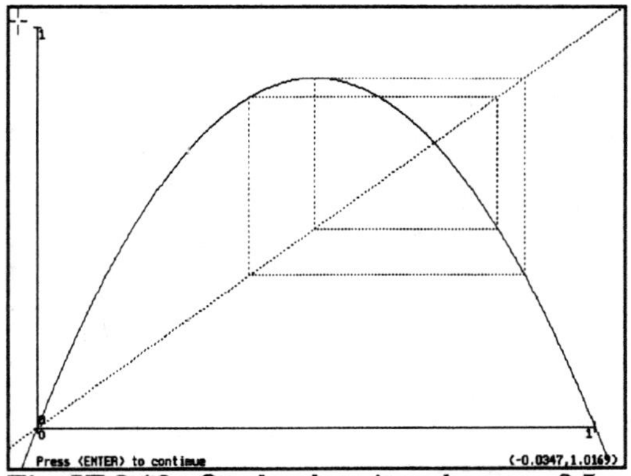

Fig. VI.2.16: Quadruple point when r = 3.5:
`Chaos2(3.5,100,30,0.6)`

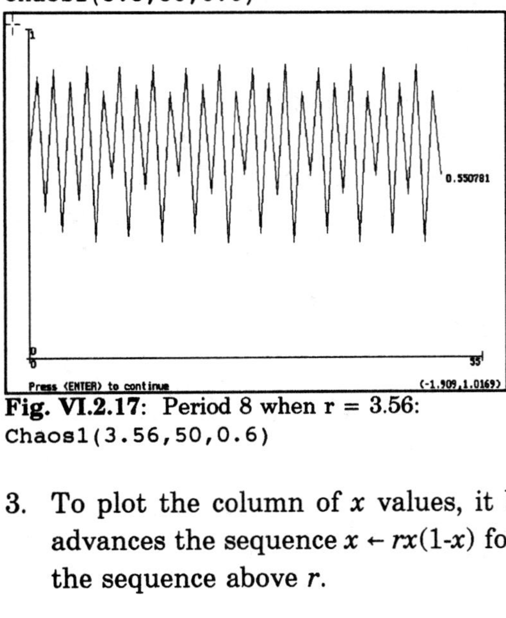

Fig. VI.2.17: Period 8 when r = 3.56:
`Chaos1(3.56,50,0.6)`

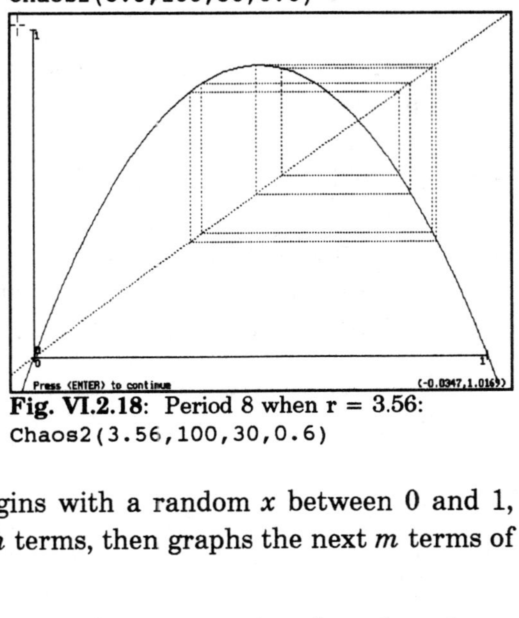

Fig. VI.2.18: Period 8 when r = 3.56:
`Chaos2(3.56,100,30,0.6)`

3. To plot the column of x values, it begins with a random x between 0 and 1, advances the sequence $x \leftarrow rx(1\text{-}x)$ for n terms, then graphs the next m terms of the sequence above r.

If the sequence of population levels is converging to a single value, the column of population values will converge quickly to a single value. If n is sufficiently large, only the final limit will be plotted. If the sequence is converging to a periodic one,

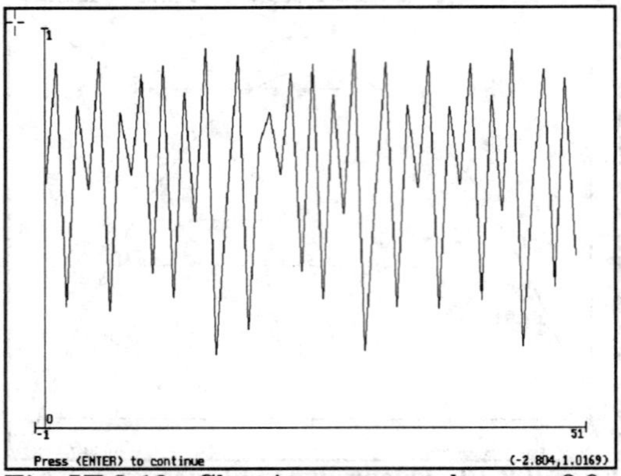

Fig. VI.2.19: Chaotic sequence when r = 3.8: `Chaos1(3.8,30,0.6)`

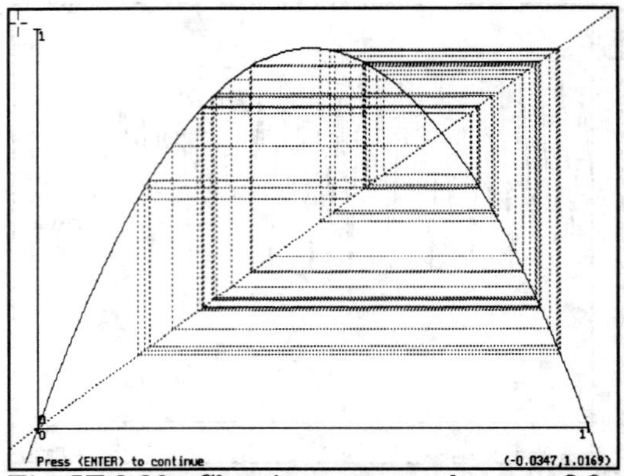

Fig. VI.2.20: Chaotic sequence when r = 3.8: `Chaos2(3.8,0,30,0.6)`

then a finite number of values will appear above r. If the sequence is chaotic, then the column of population values x above r will be spread out. Fig. VI.2.21 displays all of these characteristics. For $1 \le r \le 3$, the population converges to a single value; the column of population values x above r will be spread out. Fig. VI.2.21 displays all for $3 \le r \le 3.6$ (approximately), the population values converge to a periodic sequence, first a double sequence, then a quadruple sequence, and so forth; for $3.6 \le r$, the sequence of population values above r is chaotic. Fig. VI.2.21 was produced with the command `Bifurc(1,3,0,1,20,10,400)`.

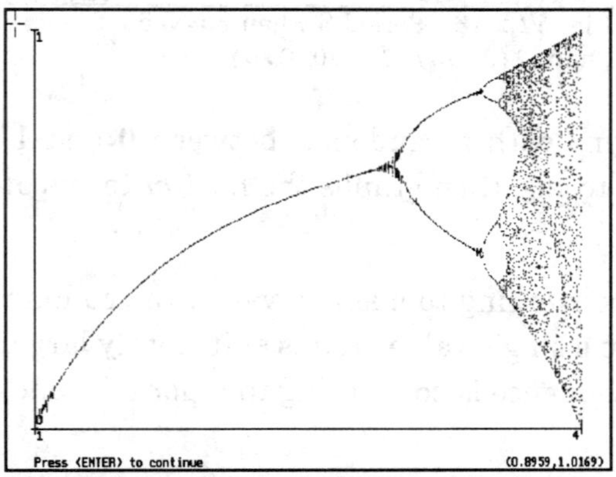

Fig. VI.2.21: Bifurcation diagram for $1 \le r \le 4$, $0 \le x \le 1$.

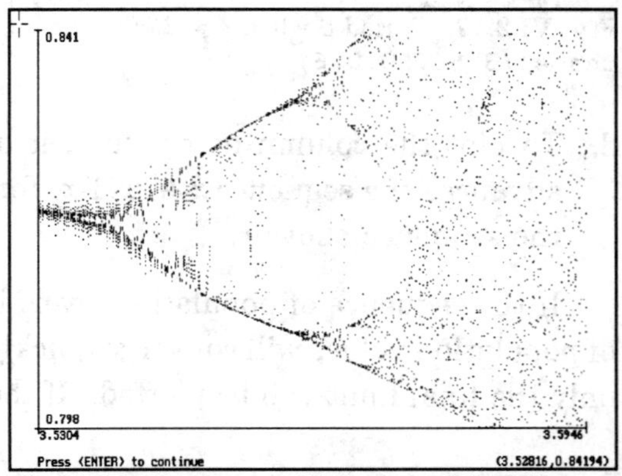

Fig. VI.2.22: A magnified part of the bifurcation diagram above.

The parameters for the subroutine allow you to magnify a part of the bifurcation diagram. Fig. VI.2.22 magnifies one of the last visible bifurcations from Fig. VI.2.21. It was produced with the command:

```
Bifurc(3.53,3.59,0.80,0.84,20,10,400)
```

This figure shows one more level of bifurcation than does the previous one. The diagrams produced here are, ideally, fractals, in that they replicate their own shapes at lower and lower levels.

```
Procedure Bifurc(a,b,c,d,n,m,q)
  // Purpose:  Show bifurcation diagram for f(x) = rx(1-x) in window
  //           (a,b) x (c,d).  The horizontal axis is the r-axis; the
  //           vertical axis the x-axis.  For each value of r, start with
  //           x = random, run n values of x, then plot the next m values
  //           of x such that c <= x <= d.  Do this for q values of r in
  //           the interval (a,b).  The plotted values of x show the
  //           stable or periodic or chaotic ranges of r.
  // Input:  a,b  :  real, the limits of the r-axis
  //         c,d  :  real, the limits of the values of x displayed
  //         n  :  positive integer, the number of values skipped before
  //                                  displaying any in each column
  //         m  :  positive integer, the number of values displayed, in
  //                                   each column,
  //         q  :  positive integer, the number of columns displayed
  f(x) = r*x*(1-x)
  Window(a,b,c,d)
  Graphics
  For r = a to b step (b-a)/q         // graph (r,x) for all these r
      x = random                      // first x value
      for i = 1 to n                  // run off n values of x
          x = f(x)
        end
      For i := 1 to m                 // plot m values of x in interval
                                      // (c,d)
          Repeat                      // next value of x in correct range
             x = f(x)
            Until x > c and x < d
          Plot(r,x)                   // plot new value of x
        end
    end
  Text
end
```

VI.2.9 Sierpinski Triangle

Problem: Construct the Sierpinski triangle by stochastic methods.

Solution: The Sierpinski triangle is the limit of the following construction: choose three points in the plane as its vertices, then pick a random point. Plot this point, then move it half way to a randomly selected vertex. Continue this process until the triangle appears. This process is called the chaos game, since the resulting triangle is a fractal curve.

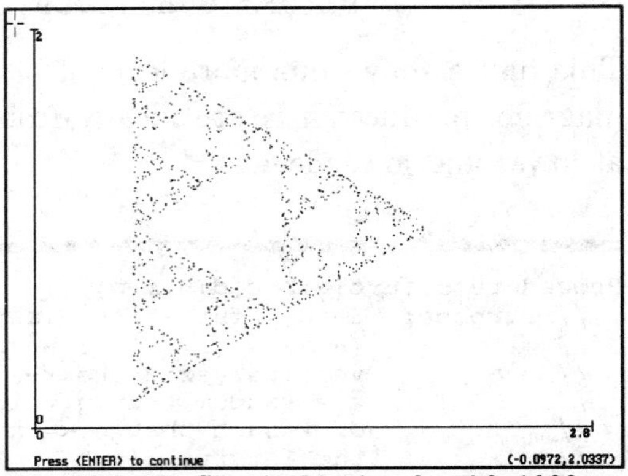

Fig. VI.2.23: Sierpinski triangle with 1000 points.

The subroutine that approximates the triangle makes use of CC's vector arithmetic. Three vertices $v[1]$, $v[2]$, and $v[3]$ are fixed, and a random point p is selected. Each of these objects is a list of length 2, so v is a list of length 3 whose entries are lists of length 2. At each step the point p is plotted, then p is replaced

```
Procedure ChaosGame(n);
  // Purpose:  choose a point at random, plot it, then at random move the
  //           point half way to one of (2, 1), (1/2, 1-√3/2),
  //           (1/2, 1+√3/2).  Repeat n times.
  // Input:  n  :  positive integer, number of points to plot

  Window(0,2.8,0,2)
  Graphics
  v = ((2,1),(1/2,1-√(3)/2),(1/2,1+√(3)/2))        // vertices of triangle
                                                   // are v[1], v[2], v[3]
  p = (2*random,2*random)                // starting point in (0,2) x (0,2)
  plot(p[1], p[2])
  For j = 1 to n
      k = random(3)                      // random integer in range 1..3
      p = (p + v[k])/2                   // move p half way to vertex k
      plot(p[1], p[2])
    end
  text                                   // hold graphic until <ENTER> pressed
end
```

by the point half-way between p and $v[i]$. A parameter passed to the subroutine by the vector sum $(p + v[i])/2$, where i is chosen at random from {1,2,3}. This controls the number of points plotted. Fig. VI.2.23 was drawn by the command:

```
ChaosGame(1000)
```

VI.2.10 Turtle Geometry and the Koch Snowflake

Problem: Develop subroutines to implement turtle geometry.

Solution: Turtle geometry (*cf* Abelson, diSissa. *Turtle Geometry*. MIT Press, Cambridge MA, 1981) is the geometry of a moving point controlled only in terms of local motion. Turtles have three characteristics:

location: a pair of (x,y)-coordinates
Heading: measured counter-clockwise in radians from due west
pen: true (down) if the turtle leaves a line when it moves, false (up) otherwise

The original turtles were remote-controlled mechanical devices that rolled about the floor on a large piece of paper. They held a pen that could be raised or lowered, so the turtle could make a drawing. Some of the terminology of turtle geometry derives from the original mechanical turtle. Turtles were invented by Seymour Papert, who also invented the computer language LOGO to control them.

In CC, a turtle is represented by a list of four values: three numbers, x,y and *heading*, and a boolean value for the pen. Controlling a turtle with purely local commands means to control the turtle by:

moving it a specified distance towards its current heading;
turning it counter-clockwise by a specified angle; or
putting the pen up or down.

Four short subroutines implement these commands. Because CC subroutines cannot use variable parameters or global variables, each of these turtle routines is written as a function subroutine. Each accepts a turtle as a parameter and returns the modified turtle. Thus, to move a turtle t a distance d, you would execute the command:

```
// Turtle Geometry workspace

// A turtle is a quadruple (x,y,h,p) where (x,y) is the position, h the
// heading (in polar coordinates: 0 = east, not north) and p the pen (true
// or false).  Since CC has no variable parameters, we use functions
instead
// of procedures to update turtle factors

Function Move(t,d)
  // Purpose:  Move turtle t distance d
  // Input:  t  :  turtle to be moved
  //         d  :  distance to move
  // Return:  Updated turtle
  a = t[1]
  b = t[2]
  t[1] = t[1]+d*cos(t[3])
  t[2] = t[2]+d*sin(t[3])
  If t[4]                              // if pen down
      line(a,b,t[1],t[2])              // draw line from old to new position
    end
  return(t)                            // return updated turtle
end

Function turn(t,a)
  // Purpose:  Turn turtle t angle a
  // Input:  t  :  turtle to turn
  //         a  :  angle to turn
  // Return:  Updated turtle
  t[3] = t[3] + a                      // change heading
  return(t)                            // return updated turtle
end

Function PenUp(t)
  // Purpose:  raise turtle's pen so it won't leave a mark
  // Input:  t  :  turtle whose pen to raise
  // Return:  Updated turtle
  t[4] = false                         // raise pen
  return(t)                            // return updated turtle
end

Function PenDown(t)
  // Purpose:  lower turtle's pen so it leaves a mark
  // Input:  t  :  turtle whose pen to lower
  // Return:  Updated turtle
  t[4] = true                          // lower pen
  return(t)                            // return updated turtle
end
```

```
t = move(t,d)
```

Turtle commands assume that CC is already in graphics mode, so they can only be used as part of subroutines. You cannot issue direct commands to a turtle.

To illustrate the uses of turtle geometry, we have developed routines that draw self-intersecting polygons and the Koch snowflake. To draw a self-intersecting polygon, you must use the *total-turning theorem*: if a turtle traces a polygon with n vertices, skipping m vertices on each step, then it goes around the origin m times and so turns a total of $2m\pi$ radians. Thus, at each vertex, the turtle turns an angle equal to $2m\pi/n$. The subroutine `Poly(n,m,new)` draws a self-intersecting polygon with n vertices, skipping m vertices at each step. If the third parameter, new, is true, then a new window is created for the drawing; otherwise the drawing is superimposed on the existing screen.

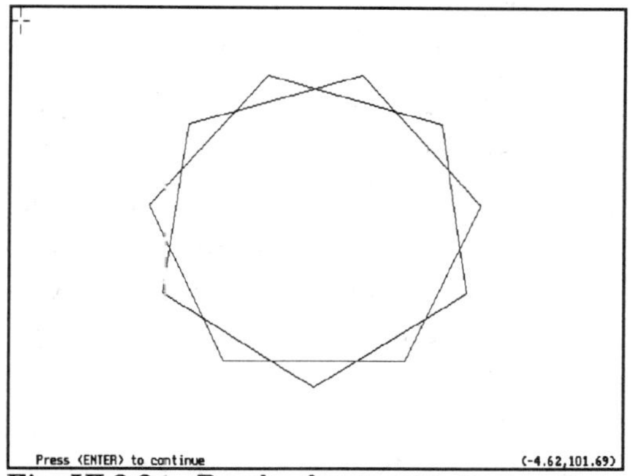

Fig. VI.2.24: Result of `Poly(11,2,true)`

```
Procedure poly(n,m,new)
  // Purpose:  use a turtle to draw a polygon with n sides skipping m
  // vertices with each side.  If new = true, start with a clean screen.
  // input:  n,m   :  positive integers.  Draw polygon with n sides,
  //                  skipping m vertices with each side
  //         new   :  boolean, if true, start with clear screen

  if new
     window(0,133,0,100)
    end
  a = 2m*pi/n                        // amount to turn at each vertex
  d = 80*cos(pi/2-a)                 // length of each side of polygon
  t = (133/2,10,a,true)              // initial turtle
  graphics
  for i = 1 to n
     t = move(t,d)                   // draw one side
     t = turn(t,a)                   // turn towards next side
    end
  text
end
```

Fig. VI.2.24 shows the result of `Poly(11,2,true)`; Fig. VI.2.25 shows the result of the following sequence of commands:

```
poly(11,2,true)
poly(11,3,false)
poly(11,4,false)
poly(11,5,false)
```

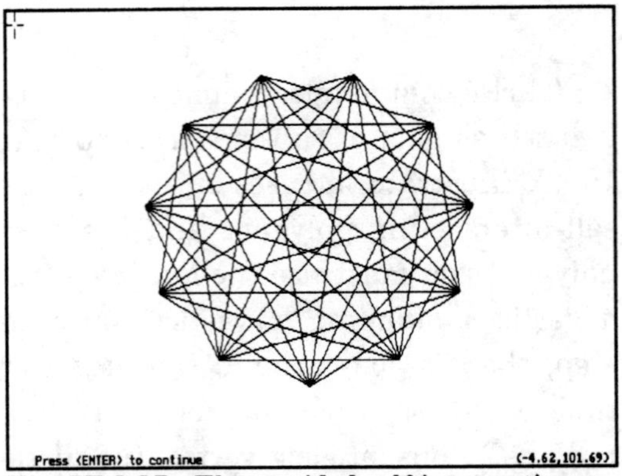

Fig. VI.2.25: Eleven-sided self-intersecting polygons

The Koch snowflake is an infinitely long, continuous, nowhere differential fractal curve bounding a finite area. Of course, we cannot draw this curve with complete accuracy; what we can draw is a sequence of curves approaching it. The subroutine *Koch(n)* draws the three sides of a Koch snowflake of level *n* using the recursive subroutine *k(t,d,n)* to draw each side. The subroutine *k(t,d,n)* moves the turtle *t* along a Koch curve of base length *d* and level *n*; this curve is made of four components, each of which is a Koch curve of base length *d*/3 and level *n*-1. The subroutine `k(t,d,n)` is written as a function that returns the updated status of the turtle *t* that it moves.

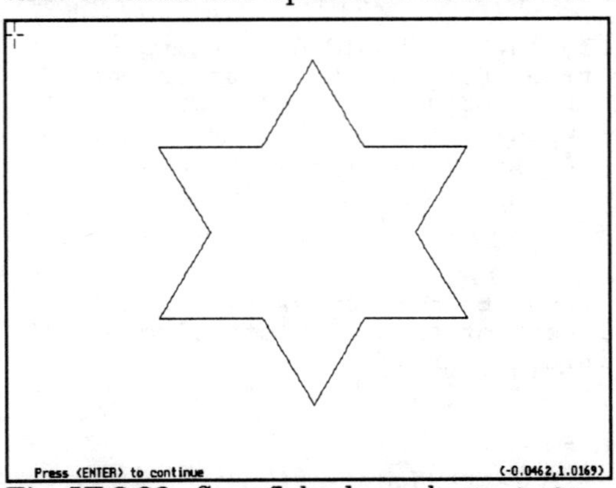

Fig. VI.2.26: Snowflake drawn by `Koch(2)`

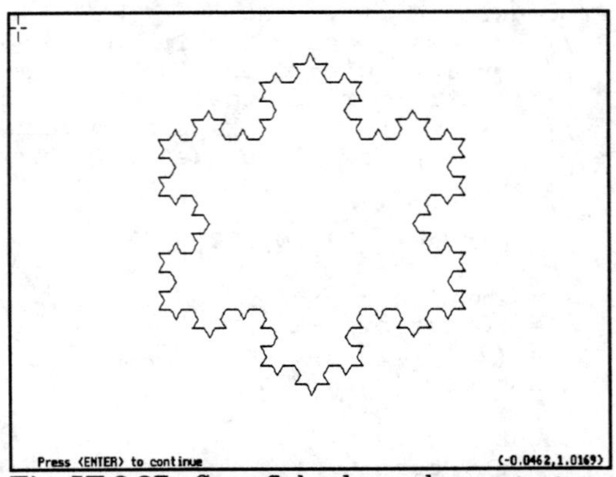

Fig. VI.2.27: Snowflake drawn by `Koch(4)`

```
Procedure Koch(n)
  // Purpose:  Draw Koch snowflake, which consists of three Koch curves
  // Input:  n  :  positive integer, degree of each component Koch curve
  window(0,1.33,0,1)
  graphics
  t = (.3,.3,pi/3,true)              // starting turtle
  for i = 1 to 3                     // draw three sides
      t = k(t,.75,n)                 // each one a Koch curve degree n
      t = turn(t,-2pi/3)             // turn toward next side
    end
  text
end

Function K(t,d,n)
  // Purpose:  Move turtle t along Koch curve base distance d of degree n.
  //           A Koch curve of degree n consists of 4 Koch curves of
  //           degree n-1, and a Koch curve of degree 1 is a straight
  //           line.
  // Input:  t  :  turtle to move along Koch curve
  //         d  :  real, total base length of curve
  //         n  :  positive integer, degree of curve
  If n == 1
      Move(t,d)                      // Move forward distance d
    else
      t = k(t,d/3,n-1)               // segment base length d/3, degree n-1
      t = turn(t,pi/3)               // turn left pi/3
      t = k(t,d/3,n-1)               // second segment base length d/3
      t = turn(t,-2pi/3)             // turn right 2pi/3
      t = k(t,d/3,n-1)               // third segment base length d/3
      t = turn(t,pi/3)               // turn left pi/3
      t = k(t,d/3,n-1)               // fourth segment base length d/3
    end
  return(t)
end
```

VI.2.11 Bar Graphs

Problem: Given a list `x[1..n]` of positive values, display the values as a bar graph.

Solution: The subroutine `Bar(x,FillStyle,HorizLabel,VertLabel)` defines an x-axis from 0 to $n+1$, a y-axis from 0 to the largest list value, and draws a rectangular bar for each list value. The base of the i'th bar goes from i-1/3 to $i+1/3$,

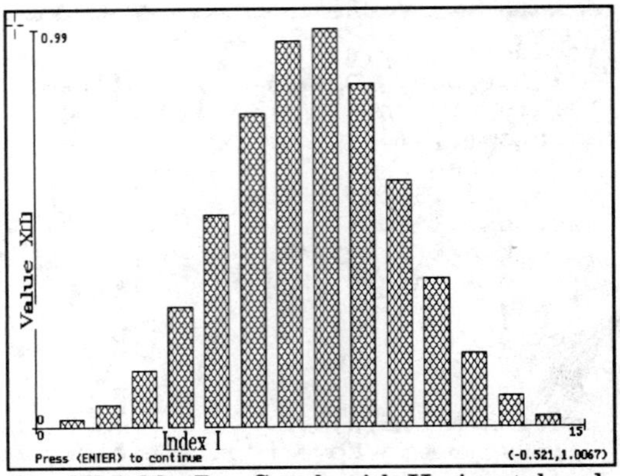

Fig. VI.2.28: Bar Graph with Horizontal and Vertical Labels

and the height is $x[i]$. The procedure also fills the bars with a pattern specified by the parameter `FillStyle` and puts labels on the x- and y-axes. Fig. VI.2.28 is a bar graph of values of the function $f(x) = e^{-x^2}$. It was constructed with the lines:

```
x = list(exp(-t^2),t=-2 to 2  step 0.3)
Bar(x,8,'Index I', 'Value X[I]')
```

```
Procedure Bar(x,FillStyle,HorizLabel,VertLabel)
  // Purpose:  Make a bar graph based on the data in the list x, filling
  //           the bars with pattern FillStyle.  Write labels on the
  //           horizontal and vertical axes.
  // Input:   x : list of positive real values to be graphed
  //          FillStyle : integer between 0 and 11, style of fill for bars
  //          HorizLabel, VertLabel : strings, used to label axes
  a = max(x)                             // highest bar
  window(0,length(x)+1,0,a)
  graphics
  VWrite@(-length(x)/180,a/4,VertLabel,2) // Write the labels
  Write@(length(x)/4,-a/18,HorizLabel,2)
  for i = 1 to length(x)                 // Draw the bars
      Rect(i-1/3,0,i+1/3,x[i])           // Outline a bar
      fill(i,x[i]/2,currentcolor,currentcolor,FillStyle)    // fill it
    end
  text
end
```

Appendix 1: Commands

Function Keys

F1	Help key	
F2	Print formulas or graphics	(III.12.1)
F3	Insert a blank line	(III.1)
Shift-F4	Delete a line	(III.1)
F5	Scroll output up	(III.3)
F6	Scroll output down	(III.3)
F7	Save formulas and graphs on disk	(III.12.3)
F8	Load formulas and graphs from disk	(III.12.3)
F9	View last graph	(III.4)
F10	Switch Input/Subroutine Windows	(IV)
Shift-F10	Clear the calculator	(III.1)
<ESC>	Quit the calculator	

Commands

^{2} (alt-2)	square	(x+3)^2
/	divide	3.43/2.97
-	subtract	8.48 - 3.26
!	factorial	17!
+	add	4.8 + 5.6
*	multiply	3.9*2.7
^	exponentiate	5.3^2.1
<	less than	y < 3
<=	less than or equal	y <= 3
<>	not equal	y <> 3
==	equals (comparison)	y == 3
>	greater than	y > 3
>=	greater than or equal	y >= 3
¦	concatenate lists	t ¦ (2,3)

Appendix I
CC Commands

π (alt-p)	same as PI	2π
Σ (alt-s)	same as SUM	Σ(j, j=1 to 10)
≤ (alt-l)	same as <=	y ≤ 3
≥ (alt-m)	same as >=	y ≥ 3
√ (alt-q)	same as SQRT	√(8)
ABS	absolute value	ABS(x-5)
ACOS	inverse cosine	ACOS(0.2)
ACOSH	inverse hyperbolic cosine	ACOSH(-4.5)
AND	logical and	(x>3) AND (x>4)
ASIN	inverse sine	ASIN(-0.1)
ASINH	inverse hyperbolic sine	ASINH(0.5)
ATAN	inverse tangent	ATAN(4.5)
ATAN2	polar angle	ATAN(3,4)
ATANH	inverse hyperbolic tangent	ATANH(0.75)
AVERAGE	average of list	AVERAGE(a)
AXISOFF	don't show axes on graphics	
AXISON	show axes on graphics	
BEEP	make a short sound	
BIN	change to binary format	BIN(27)
BINOM	binomial coefficient	BINOM(15,4)
BKCOLOR	number of graphics background color	
BLACK	number of color black	
BLUE	number of color blue	
BROWN	number of color brown	
COLOR	set color for graphing	COLOR(BROWN)
COS	cosine	COS(4.2)
COSH	hyperbolic cosine	COSH(4.2)
COT	cotangent	COT(-5.1)
CROSSX	x-coordinate of crosshairs	
CROSSY	y-coordinate of crosshairs	
CSC	cosecant	CSC(0.7+x)
CURRENTCOLOR	number of current drawing color	
CYAN	number of color cyan	
DARKGRAY	number of color dark gray	
DEGREE	change to degree mode	
DIF	differentiate	DIF(1-x^3, x=2)
DIFFG	differentiate and show graph	DIFFG(1-x^3, x=2)
DIV	integer quotient	23 DIV 7
DOTGRAPH	draw a graph with dots	DOTGRAPH(x^2,x)
DTG	same as DOTGRAPH	DTG(x^2,x)
ê (alt-e)	base of natural logs	ê^(-x^2)
ELSE	used in subroutines	ELSE IF x < 2
END	used in subroutines	
ERASE	erase graphics screen	

EVAL	immediately evaluate function definition	f(x)=EVAL(f(x-2))
EXIT	terminate subroutine	
EXP	e^x	EXP(-x^2)
FALSE	logical value	done = FALSE
FILL	fill region in graphics screen	FILL(0,0,WHITE,WHITE,1)
FIN	fast integral	FIN(SIN(x),x=0,2)
FOR	used in subroutines	FOR i = 1 to 20
FORGET	delete functions and variables	FORGET(f,g,a,b,c)
FRAC	fractional part (see INT)	FRAC(3.2)
FREQUENCIES	frequency distribution in a list	FREQUENCIES(a,0,1,0.1)
FSOLVE	fast solve	FSOLVE(cos(x)=x,x=1)
FUNCTION	start function subroutine	FUNCTION Euler(a,x,n)
GR	same as GRAPH	GR(x^2,x)
GR3	same as GRAPH3D	GR3(x^2-y^2, x=-1 to y=-1 to 1, -1 to 1, 15)
GRAPH	draw a graph	GRAPH(x^2,x)
GRAPH3D	draw a 3d-graph	GRAPH3D(x^2-y^2, x=-1 to 1, y=-1 to 1, -1 to 1, 15)
GRAPHICS	put subroutine in graphics mode	
GREEN	number of color green	
HEX	change to hexadecimal format	HEX(498)
HIDE	stop displaying a list	HIDE(a2,a4)
HISTOGRAM	draw a histogram of a list and return frequencies	HISTOGRAM(a,0,1,0.1)
î (alt-i)	imaginary square root of -1	EXP(2πî)
IF	used in subroutines	IF x < 3
IM	imaginary part (see RE)	IM(ASIN(4))
IMP	same as IMPLICIT	IMP(rx=cos(x),x=1)
IMPLICIT	define implicit function	IMPLICIT(rx=cos(x),x=1)
IN	integrate	IN(SIN(x),x=0,5)
INDEPTH	set max subdivisions for integration	INDEPTH(3)
INPUT	used in subroutines to input value from user	INPUT('next number',x)
INPUTS	used in subroutines to input string from user	INPUTS('More (Y/N)',s)
INT	integer part (see FRAC)	INT(x)/x
INTEG	integrate and graph	INTEG(sin(x),x=0,5)
INTEGER	change to integer format	INTEGER($a2d3)
INTOL	set tolerance for convergence of integration algorithm	INTOL(0.0001)
ISALIST	checks if value is a list	IF ISALIST(t)

Appendix I
CC Commands

KEEPLINE	prevent automatic color change of successive graphs	
LENGTH	length of a list	LENGTH(t)
LIGHTBLUE	number of color lightblue	
LIGHTCYAN	number of color lightcyan	
LIGHTGRAY	number of color lightgray	
LIGHTGREEN	number of color lightgreen	
LIGHTMAGENTA	number of color lightmagenta	
LIGHTRED	number of color lightred	
LINE	draw a line segment	LINE(x1,y1,x2,y2)
LINESTYLE	set style of next line	LINESTYLE(3)
LIST	define a list	LIST(t^2, t=0 to 5)
LISTG	graph two lists	LISTG(x,y)
LN	natural logarithm	LN(3.2)
LNGAMMA	log of gamma function	LNGAMMA(50.43)
MAGENTA	number of color magenta	
MARK	mark a point on the screen	MARK(x, f(x))
MAX	maximum of list	MAX(t)
MAXCOLOR	highest color number	
MIN	minimum of list	MIN(t)
MOD	integer remainder	17 MOD 3
NOT	logical operator	IF NOT ISALIST(y)
OR	logical operator	WHILE x < 3 or x > 5
PA	same as PARAMG	PA(t^2,t^3,t=-2,2)
PA3	same as PARAMG3D	PA3(r*cos(t), r*sin(t),r,r=0,1,t=0, 2π,-1,1,-1,1,-1,1,30)
PAR	parity=$(-1)^n$	Par(i)*x^i
PARAMG	parametric curve	PARAMG(t^2,t^3,t=-2,2)
PARAMG3D	3d parametric surface	PARAMG3D(r*cos(t), r*sin(t),r,r=0,1,t=0, 2π,-1,1,-1,1,-1,1,30)
PAUSE	Pause subroutine in graphics mode	
PI	π = 3.14159	SIN(5*PI/3)
PLOT	plot a single point	PLOT(x,f(x))
PO	same as POLARG	PO(sin(3Θ),Θ=0 to 2π)
POLARG	graph a polar function	POLARG(sin(3Θ),Θ=0,2π)
PRECISION	set display precision	PRECISION(7)
PROCEDURE	start procedure subroutine	PROCEDURE DISP(a,b)
PRODUCT	Product of list or terms	PRODUCT(x,x=1 to 10)
QU	same as QUICKG	QU(sin(x),x)
QUICKG	graph every 3rd pixel	QUICKG(sin(x),x)
RADIAN	change to radian mode	
RANDOM	return random number	RANDOM, RANDOM(17)
RE	real part (see IM)	RE(ASIN(4))
RECT	Draw a rectangle	RECT(x1,y1,x2,y2)

RED	number of color red	
REGR	linear regression on two lists	REGR(x,y)
REPEAT	paired with UNTIL in subroutines	
RETURN	return value from function subroutine	RETURN(x*f(x))
SEC	secant	SEC(4.5)
SHOW	display a hidden list variable	SHOW(a,b,c)
SIN	sine	SIN(-4.3)
SINH	hyperbolic sine	SINH(2.2)
SK	same as SKETCH	SK(f(x),x)
SKETCH	graph every 20th pixel	SKETCH(f(x),x)
SOLVE	solve system of equations	SOLVE(x=cos(x),x=1)
SOLVETOL	set tolerance for convergence of solving algorithm	SOLVETOL(0.0001)
SQRT	square root	SQRT(5.8)
STANDEV	standard deviation of list	STANDEV(t)
SUM	sum of list or terms	SUM(t^2,t=0,1,.1)
TABLE	display table of values	TABLE(COSH(x),x=0,1,.1)
TAN	tangent	TAN(9/8)
TANH	hyperbolic tangent	TANH(5.4)
TEXT	end graphics mode in subroutine	
TEXTCOLOR	choose colors for text screen	TEXTCOLOR(BLUE,RED, LIGHTGRAY)
THICKLINE	use thick lines for lines and graphs	
THINLINE	use thin lines for lines and graphs	
TRUE	logical value	START = TRUE
UNDEFINED	result of undefined operation	IF x == UNDEFINED
UNMARK	erase the result of MARK	UNMARK(x, f(x))
UNTIL	used in subroutines with REPEAT	UNTIL x > 5
VARYLINE	undo effect of KEEPLINE, successive graphs different colors	
VWRITE@	write vertically	VWRITE@(0,0,'y-axis')
VWRIT@C	write vertically at crosshairs	VWRITE@C('f(x)')
WHILE	used in subroutines	WHILE x > y
WHITE	number of color white	
WI	same as WINDOW	WI(0,1,0,1)
WINDOW	set graphing window	WINDOW(-1,1,-.7,.7)
WINDOWBOTTOM	y-coordinate of bottom of graphing window	
WINDOWLEFT	x-coordinate of left edge of graphing window	
WINDOWRIGHT	x-coordinate of right edge of graphing window	
WINDOWTOP	y-coordinate of top of graphing window	
WRITE	write a message box	WRITE(n,'passes')
WRITE@	write on graphics screen	WRITE@(.5,.5,'x1 = ')
WRITE@C	write at crosshairs	WRITE@C('x1 = ')
YELLOW	number of color yellow	

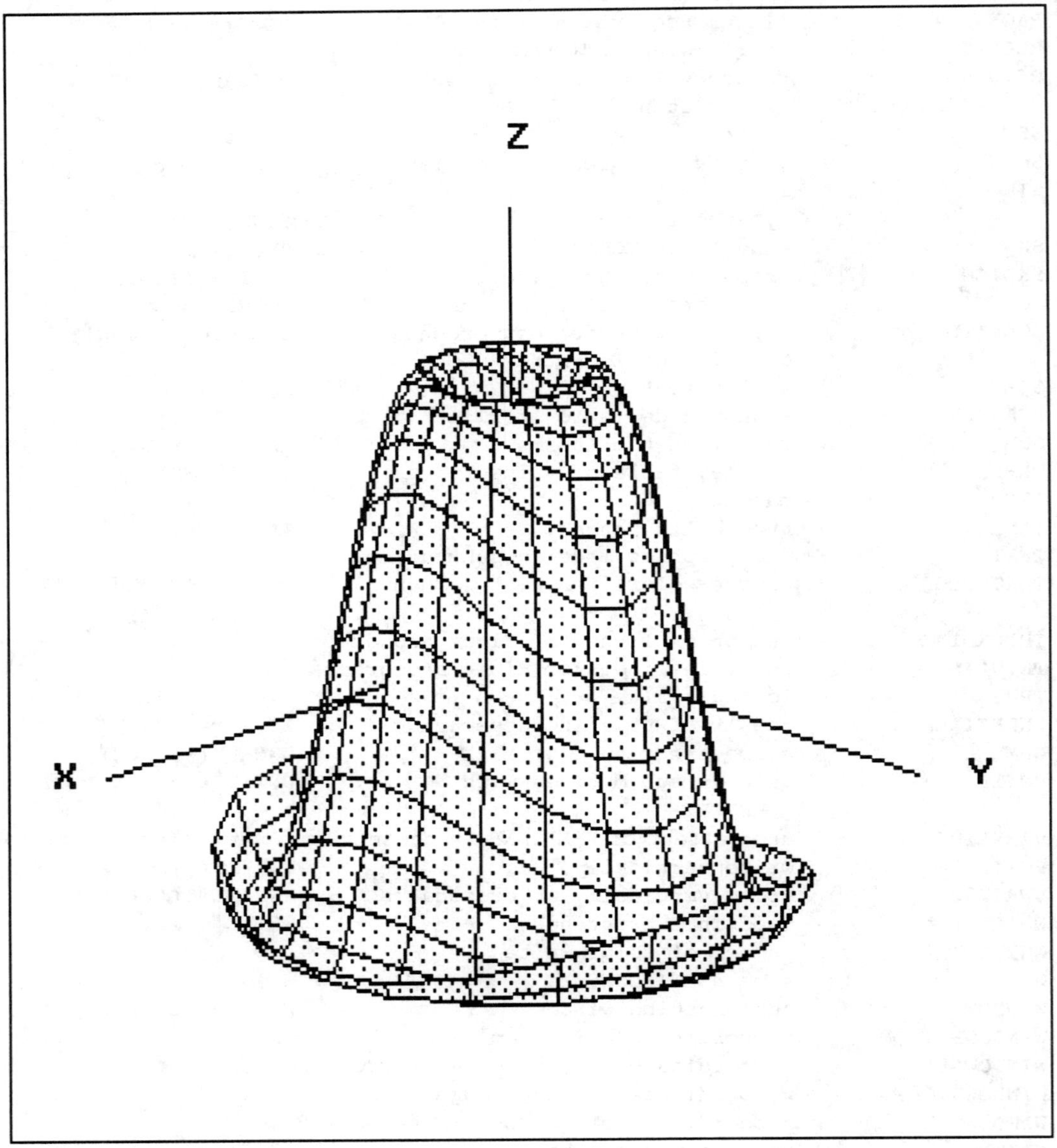

Leprechaun's Hat

Appendix 2: Error Messages

`<command> cannot be applied to a string`
: You have used a command like `SIN` on a string value instead of a number.

`<command> expects arguments`
: Many CC commands require arguments. For example, you cannot enter `SQRT` by itself.

`<command> unexpected`
: Line containing `PROCEDURE` or `FUNCTION` or `EXIT` appears in unexpected place in a subroutine.

`<formula> appears where variable name expected`
: Sometimes a simple variable name is required, not a formula. For example, when you enter `IN(f(x),x=a to b)`, `f(x)`, a and b can be formulas, but x must be a variable name.

`<formula> includes undefined or non-integral or negative parameter`
: Only a positive integer can be used as an element selector for lists.

`<formula> is not a list`
: In a sublist or element selection `x[a]`, x is not a list. You might have entered `x[a,b]`, and `x[a]` was not a list.

`<formula> is not a variable name`
: Sometimes a simple variable name is required, not a formula. For example, when you enter `IN(f(x),x=a to b)`, `f(x)`, a and b can be formulas, but x must be a variable name.

`<formula> is not a valid number`
: Invalid numeric format was used.

`<formula> is out of range`
: Sublist selector picks out an element beyond the end of the list.

`<formula> is undefined`
: Sublist selector is undefined value.

`<formula> must be a single list entry, not a range`
: You cannot enter: `X[3:5] = (2,3,4)`.

`<formula> out of range or undefined`
: Sublist selector too large or undefined.

`<input line> has incorrect parentheses`
: Unbalanced parentheses discovered.

Appendix II
Error Messages

<name> cannot be a function name
<name> cannot be a variable name

Probably uses illegal characters.

<name> is a built-in operator

You tried to used a pre-defined name like SIN as the name of one of your variables or user-defined functions or subroutines.

<name> is a variable, not a function

You used a variable name x in the function construction x(...).

<name> is already a variable or function name

So it cannot be used to name a subroutine.

<name> is not a list

In trying to construct a sublist x[t], x is not a list.

This error message also appears if you try to HIDE a variable that is not a list.

<name> is the name of a pre-defined operation
<name> is the name of a scratchpad procedure
<name> is the name of one of your functions
<name> is the name of one of your variables

When you define a function or variable or subroutine, you cannot use a name that has already been used for a built-in function or variable or function or subroutine.

<name> is not a graphics file

The file name you specified was not a graphics file.

3d graphs cannot be drawn while subroutine in GRAPHICS mode

A subroutine tried to draw a 3d-graph after executing the command GRAPHICS.

ANS cannot be hidden

The command HIDE cannot apply to the variable ANS.

Arguments after RANDOM not correct

In RANDOM(n), n must be a positive integer.

Arguments are not colors or color numbers

In the command COLOR(n), n must be either a color word or a number between 0 and a system-dependent maximum MAXCOLOR.

Background color must be different from foreground color

In the command TEXTCOLOR, the two foreground colors must be different from the background color.

background color must be from first half of color list

In the command TEXTCOLOR, not all colors can be used as background colors. Background colors are the darker colors.

Beginning sublist limit too big

In a sublist specification x[m:n], m is too large for x.

Appendix II
Error Messages

```
BUG -- null op code encountered
BUG--illegal value sent to OutputString
BUG -- Variable Stack Underflow
BUG -- empty control stack
BUG - LastLocalVar < 0
BUG -- Cannot Assign value to <formula>
BUG -- attempt to execute Nil code pointer
BUG--<name> was made into a name
BUG--LastError < 0
```

None of these error messages should ever appear. They indicate an internal inconsistency in CC. If you can make one of them appear consistently, please communicate with the author. A free correction will be given to the first person who reports the error.

`Cannot differentiate <formula>`

Because the formula includes a function that cannot be differentiated, like `MOD` or `LNGAMMA`.

```
Cannot differentiate PRODUCT with non-constant limits
Cannot differentiate SUM with non-constant limits
```

The function `PRODUCT(f(x),x=a to b step c)` and `SUM(..)` cannot be differentiated unless `a`, `b` and `c` evaluate to constants.

`Cannot interpret <expression>`

All-purpose error message when a line cannot be interpreted.

`Cannot interpret <expression> ...`

The part of the line beginning with <expression> cannot be understood.

`Cannot interpret <input line>. Should = be ==?`

This error usually occurs when you enter `x = y` when you needed to use `x == y`. It indicates that an equals sign was found where it could not be interpreted as a variable assignment or function definition.

`Cannot interpret formula.  Possibly larger than max real`

CC thinks this formula is a number which is too large to evaluate.

`Cannot load last graph from disk`

CC cannot read the last graph file from the disk where it was written. Perhaps you wrote to a floppy and removed it from the computer.

`Cannot set the color of a monochrome screen`

The command `TEXTCOLOR` cannot be used on a monochrome system.

`Cannot understand variables in <subroutine heading>`

The variable names used in a subroutine heading are not correct.

`Cannot use <command> in calculator mode`

Commands like `REPEAT` and `GRAPHICS` cannot be used except in subroutines.

Appendix II
Error Messages

`Cannot write table while in graphics mode`

In a subroutine, you cannot execute the command `TABLE` if you previously executed `GRAPHICS.` Tables cannot be written to the graphics screen.

`Color code must be between 0 and <system-determined constant>`

Most color systems have 16 colors numbered 0..15. The color code must evaluate to one of these values.

`Command GRAPHICS can be used only on the scratchpad`

You cannot use the command `GRAPHICS` in the Input Window.

`Computing interrupted--press any key to continue`

When you press <ESC> while a command is executing, you generate this error.

`Could not write last graph to disk. Check for full disk. Try again (Y/N)?`

The disk where the last graph was to be written is not available or full. If it is a full or defective floppy, replace it with a blank, formatted floppy and try again.

`DiffG requires 3 parameters`

The command `DIFFG(f(x),x=a)` has three parameters `f(x)`, `x`, and `a`.

`ELSE IF not followed by boolean value`

In a subroutine, an `ELSE IF` clause must be terminated by a formula that has a boolean value `TRUE` or `FALSE.`

`Emphasized foreground color is not correct or is the same as the foreground color`

When you execute `TEXTCOLOR(foreground, emphasized, background)`, the emphasis color must be a legal color different from the foreground color.

`Empty sublist specification`

You tried to execute a sublist specifier `x[]`.

`Empty variable name encountered`

Somehow a variable name of length 0, '', was generated.

`Error after SHOW or HIDE`

The parameters of `SHOW` or `HIDE` are incorrect. Probably one of them is not a legal variable name.

`Error in: <input line>`

CC could not interpret this line and is not sure why.

`Error in arguments for SOLVE or IMPLICIT`

Perhaps you entered `Solve(F(x),x=a)` instead of `Solve(F(x)=0,x=a)`.

`Error in sublist specification`

See the manual for a description of the correct arguments.

`FOR must be followed by indexing variable and 2 or 3 parameters`

Error in arguments following FOR.

`Foreground color is not correct`
: The color specified is not between 0 and the highest legal color number, usually 15.

`FORGET can only be applied to a variable or function`
: You probably tried to `FORGET` a formula or an undefined term.

`Formula is too complex to evaluate--check for infinite recursion`
: Usually this happens when you have made a recursion error, but if you are short of memory and you try to evaluate a complex formula, you may run out of memory and generate this error.

`Function <name> requires <number> arguments`
`Function <name> requires <number> variable(s)`
: You tried to use one of your own user-defined functions with the wrong number of variables.

`Function argument <name> is not a variable name`
: In defining a function `f(x) = ...`, the expression `x` is not a legal variable name.

`Graphs can be printed only on Epson, IBM or HP-compatible printers`
: If you declared your printer type as "none of the above", then you can only write text, not graphics to it.

`Halting because of insufficient memory`
: This unusual message occurs when writing error messages to the graphics screen and running out of memory. CC cannot recover from this error.

`histogram cannot be drawn while subroutine in GRAPHICS mode`
: You cannot use `HISTOGRAM` after `GRAPHICS` in a subroutine.

`IF or WHILE or FOR does not have matching end`
: CC was interpreting a subroutine, and was in an `IF` or `WHILE` or `FOR` clause when it came to the end of the subroutine, usually by coming to a line beginning `PROCEDURE` or `FUNCTION`.

`Illegal decimal point in a number`
: Usually this means two decimal points in a number.

`IMPLICIT has incorrect arguments`
: See the manual for a description of the correct arguments.

`IMPLICIT has too many arguments to be differentiated`
: You can define an implicit function with up to eight equations, but you cannot differentiate it if it has more than four equations.

`Incorrect arguments for HISTOGRAM`
: See manual for correct arguments.

`Incorrect limits for graphing or too many points requested`
: The limits for `PARAMG` or `POLARG` are incorrect (usually the last limit is lower than the first) or more than 1000 points are specified.

`Incorrect number of variables for <subroutine name>`
: You called a subroutine from the Input Window or from another subroutine with the wrong number of variables.

`Incorrect parameters for <function definition>`
: In defining a user-defined function, `f(x,y,...) = ...`, one of the variables is not a legal variable name.

`Incorrect parameters for Fill`
`Incorrect parameters for TABLE`
: See manual for correct parameters.

`Indexing variable expected in FOR statement`
: The first parameter after `FOR` must be a variable name.

`Input symbol <formula> is longer than 34 characters`
: Usually this means that you have mistyped something, so CC could not parse your input line into separate formulas.

`Insufficient memory for 3d graphing`
: See the manual for hints on increasing available memory.

`Insufficient memory to store 2d graph`
: You tried to transfer the last graph from disk to memory, and there isn't enough memory to hold it.

`Insufficient Memory to save or print graph`
: Saving the printing a graph takes a bit of memory. It cannot be done when less than 5000 bytes are available.

`Integration depth must be a positive integer < 10`
: To go deeper than this would be too slow and pointless.

`Integration tolerance must be a real number`
: It must also be positive.

`Last 2d graph not found on disk`
: Maybe you saved the last graph on floppy and then removed the floppy from your computer.

`Limits for GRAPH3D or PARAMG3D not correct`
: Usually this means that one of the beginning limits is larger than its corresponding ending limit, or one of the limits is undefined.

`Line style must be between 0 and 3`
: These are the only legal style numbers.

`List did not contain any real data for histogram`
: Probably the list did not evaluate because of some undefined term.

`LISTG must be applied to a pair of non-empty lists`
: Check that both arguments evaluate to lists.

`Local variable expected after <formula>`
`Local variable <name> is not a variable name`

Usually this means that you forgot to put in the local variable `x` in a command like `GRAPH(f(x),x)` or `DIF(f(x),x=a)`.

`Magnification factor in Write@ must be an integer between 1 and 10`

In the command `WRITE@(x,y,message,size)`, `size` must be in the range 0..10.

`Maximum number of subdivisions is 40`

In the commands `GRAPH3D` and `PARAMG3D`, the last parameter must be no more than 40.

`Message in Write@ starts off the graphics screen`

In the command `WRITE@(x,y,...)`, the point `(x,y)` is off the graphics screen.

`Message in Write@ would extend off the graphics screen`

Actually, `Write@` puts characters on the screen one at a time. You get this error when the characters start to run off the screen.

`Missing parenthesis in <input line>`

Usually this means unbalanced parentheses.

`Name <name> is longer than 12 characters`
`Name <name> must begin with a letter`

Variable and function names cannot be longer than 12 characters and must begin with a letter.

`Not enough memory to complete this operation`

So we have to go back and start over. See the manual for information about saving memory.

`Not enough memory to write message to graphics Screen--Halting`

This happens at a point when CC cannot even stop itself and go on, so it must halt.

`Number <formula> is larger than maximum real:  <maximum real>`
`Number <formula> is too large`

The largest real is about 10^{37} for CC and 10^{4000} for CC87.

`Only IF ... can follow ELSE`

Perhaps you wrote `ELSE x = 5`. The command after `ELSE` must be on its own line.

`Only numbers and strings can be compared`

You cannot compare a number to a string or anything to a list.

`Parameter %<name> not matched by function name`

You called a subroutine which had a function name parameter, and you did not use the name of a function to match the parameter.

Appendix II
Error Messages

`Parameters for WINDOW not correct`

Usually this means that the lower limits are greater than the upper limits, or one of the limits is undefined.

`Parameters in <line> return incorrect value`

Usually this means that the expression after `WHILE` or `UNTIL` or `IF` or `ELSE` did not return a boolean value `TRUE` or `FALSE`. Perhaps the value was `UNDEFINED`.

`Parameters in RETURN(...) cannot return a value`

The formula in a return statement is not a function; it might be a graphing command or some other command that does not produce a value.

`possible unbalanced parentheses or apostrophes: <input line>`

When interpreting a formula, CC first counts the parentheses and apostrophes. This error message means that it found them out of balance.

`Precision value must be an integer`
`Precision value must be between 0 and 10 (15 for CC87)`

In the command `PRECISION(n)`, n must be an integer in the specified range.

`Printer not working. Try again (Y/N)?`

`Check if your printer is turned on, reset, and the paper is ready.`

`Printing -- if no response, check printer`

This message shows while you are printing. So long as your printer is producing, there is nothing to worry about. Take action if your printer isn't printing.

`Random must have an integer argument 1..65535 or none`

`RANDOM` returns a random real in the range (0,1); `RANDOM(n)` returns a random integer in the range `1..n`. The number n must be between 1 and $65535 = 2^{16}$-1.

`Second sublist limit is too big`

In the sublist expression `x[a..b]`, `b` is larger than the length of `x`.

`SOLVE and IMPLICIT can use no more than 8 equations`

When you solve a system of equations, you must have the same number of equations as variables and no more than eight of each.

`Sublist limits not correct`

In the expression `x[a:b]`, either `a` or `b` cannot be interpreted as an integer.

`TABLE is limited to 20 lines`

In the command `TABLE(f(x),x=a to b step t)` the number of lines $1+(b-a)/t$ cannot be more than 20.

`The command <command> must be the only command on a line`

Commands like `GRAPH` and `TEXTCOLOR` must be the only command on a line.

`The function <name> includes undefined parameters`

When you call a subroutine with a function name parameter, the function you pass to the subroutine must be completely defined. For example, if `f(x) = x+a` is a function definition, and you use `f` as a parameter in a function call: `MidPt(f,1,3,10)`, then the constant a must be defined or you will get this error. You cannot pass a function name to a subroutine if the function has undefined components.

`The reserved word <name> cannot be a function name`
`The reserved word <name> cannot be a variable name`

You tried to redefine one of CC's built-in functions.

`Tolerance for solving algorithm must be a real number`

The parameter `t` in `SolveTol(t)` must be a positive real number.

`Too complex: <input line>`

The line you have entered is too long to be interpreted.

`Too many function variables stored`

There are too many variables in the function you are trying to define.

`Too many local variables`
`Too many local variables: <input line>`

You have an expression that uses too many local variables, which are the variables `x` in expressions like `Graph(f(x),x)` and `Dif(f(x),x=3)`.

`Too many messages--halting`

`Too many screen messages printed before prior ones cleared. This is a fatal error.`

`Unbalanced apostrophes found`

Your input line includes an odd number of apostrophes. Perhaps you forgot to double one of the apostrophes in a string like `'x''='`, which means x'=.

`Unbalanced parentheses: <input line>`

CC checks each line for balanced parentheses.

`Unexpected character in numeric expression <formula>`

This is an unusual error, since a letter in a number usually makes CC think the number has ended and a variable started.

`Unexpected END in REPEAT loop`

`REPEAT` loops end with `UNTIL...`, so an `END` is an error.

`Unexpected end to <subroutine name>`

While executing a subroutine, CC encountered a line beginning `FUNCTION` or `PROCEDURE`. CC thinks it came to the end of one subroutine and the beginning of the next one, but there was no `END` to end the first subroutine.

`Unrecognized name <name> after FORGET`

Told to `FORGET` a variable, CC cannot find the variable to forget.

`UNTIL without matching REPEAT`

CC encountered a line with `UNTIL`, but it had not entered a `REPEAT` loop.

Appendix II
Error Messages

`Use EVAL only at the beginning of a definition`

The correct syntax is `f(x) = EVAL(...expression using x...)`. You cannot write `f(x) = 3 + EVAL(g(x))`.

`Value following ELSE is UNDEFINED`
`Value following WHILE is not boolean`
`Value following WHILE is UNDEFINED`

`ELSE` and `WHILE` must be followed by an expression that evaluates to `TRUE` or `FALSE`.

`Value is too large to take integer or fraction part`

The functions `INT` and `FRAC` can be applied only to values between -2,147,483,648 and 2,17,483,647.

`Variable expected in INPUT statement`

The message is optional in `INPUT(<message>, variable)`, but the variable to receive the input must be specified.

`Wrong number of arguments following <command>`

You tried to use a built-in function with the wrong number of variables.

`Wrong number of arguments for TABLE`

`TABLE` requires one or two functions, a local variable, beginning and ending parameters and step size. The step size cannot be omitted.

`Wrong number of variables for a function`

You tried to use a user-defined function with the wrong number of variables.

`You cannot FORGET the variable ANS`

You can erase any other variable, but not the default variable `ANS`.

INDEX